Brandon Clark

Types of Boilers

Instruction Paper

By
Carl S. Dow, S.B. (Harvard University),
American Society Mechanical Engineers

Deaerator Series Section by
Shippensburg Pump Co., Inc. (SHIPCO®)

BURD STREET PRESS
SHIPPENSBURG, PENNSYLVANIA

Prepared by Carl S. Dow, S.B. (Harvard University),
American Society Mechanical Engineers

This Burd Street Press publication
was printed by
Beidel Printing House, Inc.
63 West Burd Street
Shippensburg, PA 17257-0708 USA

The acid-free paper used in this book meets the guidelines for permanence and durability of the Committee on Production Guidelines for Book Longevity of the Council on Library Resources.

For a complete list of available publications
please write
Burd Street Press
Division of White Mane Publishing Company, Inc.
P.O. Box 708
Shippensburg, PA 17257-0708 USA

Library of Congress Cataloging-in-Publication Data

Dow, Carl S. (Carl Stephen)
Types of boilers : instruction paper / by Carl S. Dow. Deaerator series section / by Shippensburg Pump Co., Inc. (SHIPCO)
p. cm.
ISBN 1-57249-382-8 (alk. paper)
1. Steam-boilers--Handbooks, manuals, etc. I. Title: Deaerator series section. II. Shippensburg Pump Co. III. Title.

TJ289.D69 2006
577.7'27--dc22

2005058274

PRINTED IN THE UNITED STATES OF AMERICA

Contents

Types of Boilers

Generally speaking, a steam boiler is a closed metallic vessel in which steam is generated from water by the application of heat. As steam is under pressure it is evident that the vessel must be strong and tight.

To operate the boiler safely and economically there must be certain fittings and accessories—some of these are used in the care of the boiler, while others serve to increase the economy. Among the most important attachments and appurtenances may be mentioned the following:

A feed pump or injector, with valves, piping, etc., to supply water to the boiler.

Gage cocks and glass water gage to show the attendant the height of water or the water level, as it is called, in the boiler.

A pressure gage to show the pressure of steam in the boiler. The pressure is usually measured in pounds per square inch.

A safety valve to allow steam to escape from the boiler when the pressure exceeds a certain fixed amount. This attachment, being a safety device, should be automatic and reliable.

A blow-off pipe, with its valves, to blow out sediment from the boiler, reduce the amount of water in the boiler, or empty it.

A steam pipe, with its valves, to conduct the steam from the boiler to the place where it is to be used.

Manholes and handholes, with covers, for examination, repairs, and cleaning.

[*]**Fusible plugs** to give warning when the water level becomes too low, or melt and allow the water to escape.

[*]**High- and low-water alarms** to give warning when the water level is too high or too low.

[*]**A heater** to raise the temperature of the feed water as nearly as possible to that of the water in the boiler.

In addition to these there are other attachments such as:

Lugs or brackets for supporting the boiler.

Masonry for setting the boiler and keeping it in position, and in many cases to keep the hot gases in contact with the shell.

Furnace fittings, including grate bars, bearer bars, dampers, fire doors, ashpit doors, etc.

The chimney to carry away the waste gases and create draft.

Tools, such as shovels, slice bars, scrapers, tube brushes, etc.

[*] Although the last three are desirable, they are not absolutely necessary, as a boiler can be successfully operated without them.

Definitions

The following definitions should be remembered in connection with the terms used in designating the various classes.

A fire-tube boiler is one having the heating surface composed largely of tubes which are surrounded with water, the hot gases passing through them.

A water-tube boiler is also composed of tubes, but in this case *water* flows through the tubes, while the hot gases pass around and among them.

In a sectional boiler the tubes and corresponding headers form comparatively small units. Each unit is complete in itself; that is, it is in communication with a steam and water drum but is independent of the other units.

A non-sectional boiler is one having all the tubes in communication with one another; in other words, all or nearly all the tubes are expanded into a common header or drum. The boiler is not made up of units.

A single-tube boiler is made up of plain tubes.

A double-tube boiler has a small tube inside of the regular tube and concentric with it.

A boiler is **externally-fired** when the furnace is separate from the shell; in such boilers the fire is usually placed in a brick furnace.

In the **internally-fired** boiler the grate is inside of a flue which is within the shell.

A fire-box boiler is one having the fire within a fire box which, although external to the shell, is rigidly connected to it. The fire box is usually made of steel plates instead of brick as in the case of the externally-fired boiler.

Classifications

The almost endless variety of boilers now in use is due largely to the many conditions under which they are used. Other reasons for the numerous forms are the great latitude in design and construction, and the competition among engineers, who have, during the last century, sought to produce, at moderate cost, steam generators that will be safe, durable, and economical.

The necessity for careful classification before discussing the details is apparent when one considers the similarities and differences. Much valuable time may be saved by selecting some make of boiler to represent a given class. Still further, the classification reduces the chances of overlooking interesting features.

CLASSIFICATION.

According to Use.

- **Stationary**
  - Early Forms.
  - Plain Cylindrical
  - Single Flue, Externally-fired
  - Flue Boilers
    - Cornish (single-flue)
    - Lancashire (two-flue)
    - Galloway
  - Multitubular
    - Externally-fired
    - Internally-fired (return-tube)
  - Fire-box Boilers
    - Horizontal
    - Vertical
  - Water-tube Boilers
    - Straight-tube
    - Curved-tube
    - Horizontal
    - Vertical
    - Sectional
    - Non-sectional
  - Mixed Types
  - Peculiar Forms

CLASSIFICATION.

According to Use.

- **Marine**
  - **Early Forms** (box or rectangular)
  - **Scotch or Drum**
  - **Return-tube**
  - **Through-tube**
  - Water-tube
    - **Curved-tube**
    - **Straight-tube**
    - **Sectional**
    - **Non-sectional**
  - **Launch Boilers**
- **Locomotive**
  - **Multitubular** fire-box (common form)
  - **Wooten Type**
  - **Corrugated Furnace**
  - **Peculiar Forms**

Boilers may be classified in many ways. They may be divided into the following great classes: Fire-tube and water-tube, vertical and horizontal, stationary and non-stationary, or externally-fired and internally-fired. They may also be classified according to uses or according to forms of construction. For illustration, two classifications, of which the following seems better for this discussion, are given.

CLASSIFICATION.

According to Form of Construction.

Early Forms.

- **Flue**
  - Cornish (single-flue)
  - Lancashire (two-flue)
  - Galloway
  - Single Flue (externally-fired)
- **Fire-tube** (Multitubular)
  - Horizontal (common form)
  - Vertical
  - Return-tube
  - Through-tube
  - Fire-box
  - Peculiar Forms
- **Water-tube**
  - Horizontal
    - **Straight-tube**
      - **Sectional**
      - **Non-sectional**
    - **Curved-tube**
  - Vertical
    - **Straight-tube**
    - **Curved-tube**
  - Peculiar Forms

Mixed Types.

Chapter 1

Early Forms

The earliest boilers of which we have reliable record were spherical. They were of cast iron and set in brickwork. It was customary to set this type of boiler with the fire underneath and construct flues in the brickwork to conduct the hot gases around the boiler just below the water level. The hot gases passed entirely around the boiler before escaping to the chimney.

The Haystack Boiler. The next form to be generally used was that invented by Newcomen in 1711. On account of its peculiar shape it was called the "Haystack" or "Balloon" boiler. It was of wrought iron and had a hemispherical top and arched bottom. The fire was placed underneath the arched portion; the hot gases surrounding the lower part of the boiler. An improved form of the Haystack boiler is shown in Fig. 1. Smeaton placed the fire inside the shell and arranged internal flues for conducting the hot gases to the

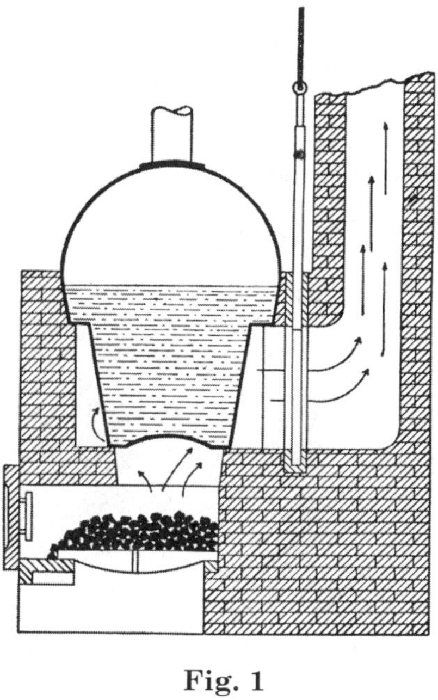

Fig. 1

chimney. This arrangement increases the heating surface and consequently the economy of the boiler.

The Wagon Boiler. To still further increase the heating surface, James Watt introduced his "Wagon" boiler. This form is shown in Fig. 2. The top was cylindrical and the sides curved inward. The curved plates assisted in the formation of flues on either side. The hot gases passed from the grate, underneath the boiler to the rear, through the left-hand flue to the front, then through the right-hand flue to the rear and thence to the chimney. This was called the

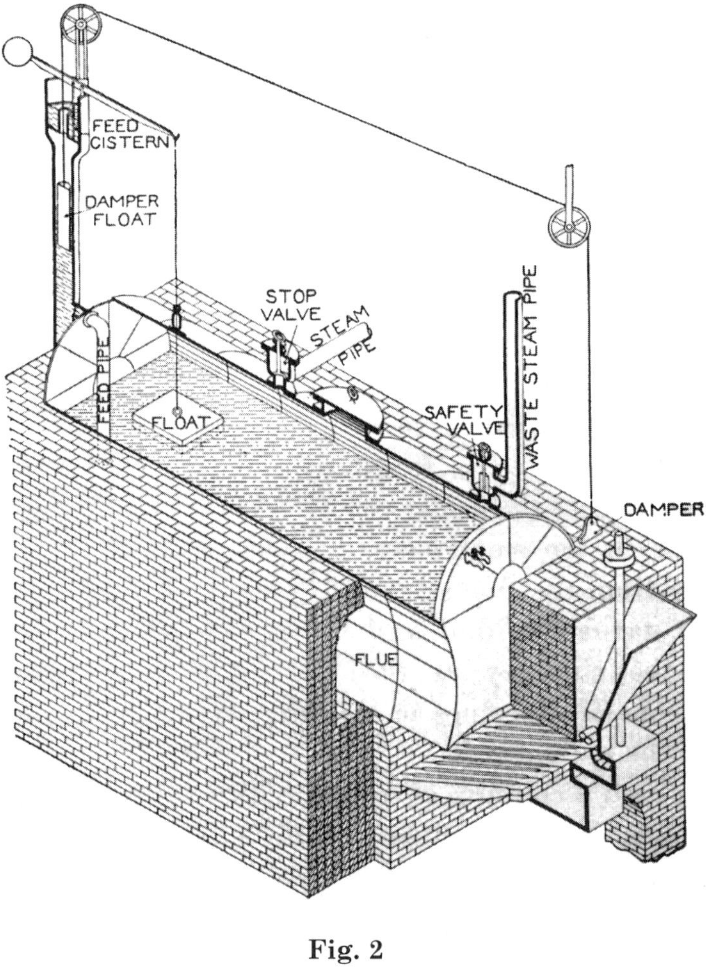

Fig. 2

wheel draft because the gases passed entirely around the boiler. In the large sizes a flue was placed in the boiler. The products of combustion returned through this flue to the front after passing under the boiler to the rear, as in the small sizes. On issuing from the flue at the front, the gases divided and passed to the chimney at the rear by means of the flues in the brickwork. This form of draft was called the *split draft*.

Watt used a column of water in the vertical feed pipe as a pressure gage; the rise and fall of this column also controlled the damper. The feed was regulated by a float.

Modern Boilers

Although such boilers as the Haystack, Wagon, and others were fairly satisfactory in the period in which they were invented, they could not stand the higher pressures that soon became common.

About the beginning of the nineteenth century the cylindrical boiler was introduced. The earliest forms were the plain cylindrical boiler and the "Egg-end" boiler. The difference was the form of the ends—those of the former were flat and of cast iron, while the ends of the latter were hemispherical and made of wrought iron. The egg-end boiler required no staying or bracing because its form is, with the exception of a sphere, the strongest to resist internal pressure.

The Cylindrical Boiler consists of a shell of wrought-iron boiler plate and ends of the same material or of cast iron. It was set in brickwork as shown in Fig. 3. The boiler was about two-thirds filled with water, the remaining third forming the steam space. The collect and store the steam as it rose from the water a steam dome was added. The steam pipe was attached to the dome to which the safety valve also was connected. The hot gases from the fire passed under the boiler to the rear and then to the chimney.

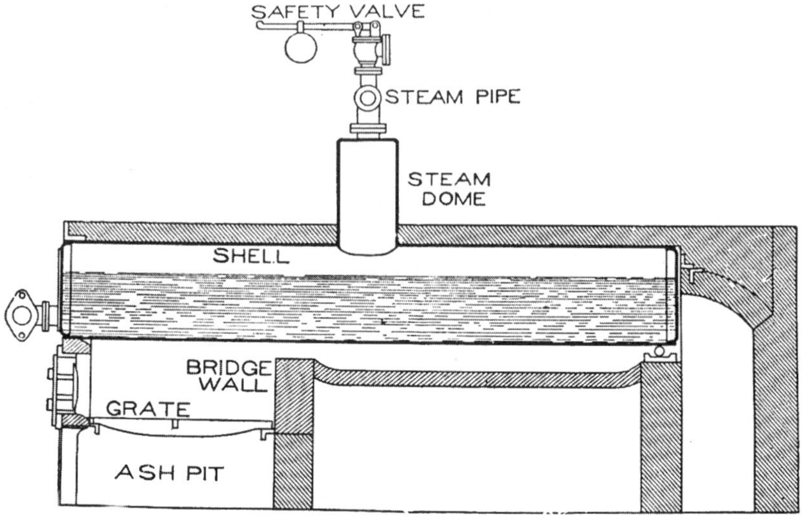

Fig. 3

The heating surface of this type is small with a given diameter unless the boiler is made very long. As all sediment collects in the bottom, where the heat is most intense, the plates are liable to burn. Since sediment and scale are poor conductors of heat, the heat remains in the plates and overheats them instead of flowing to the water.

The disadvantages (the small heating surface and the collection of sediment) do not seem so serious when one considers the simplicity of construction, strength, durability, and ease of repairing and cleaning.

The plain cylindrical boiler was adapted for mining districts, iron works and other places where fuel is abundant and skilled boiler makers are not readily found. This boiler was made very long to get the required heating surface, the length sometimes exceeding fifty feet.

Chapter 2

Flue Boilers

In order to get the necessary heating surface in the cylindrical boiler without making it excessively long, it was made with an internal flue through which the hot gases passed to the chimney. This flue was quite large and extended from end to end. In the United States, Oliver Evans used this type in 1800. In England, it led to the internally-fired flue boilers which were so extensively used.

The Cornish Boiler

Horizontal—Single-Flue—Internally-Fired

When it was found that about 25 percent of the total heat of combustion was lost by radiation from the furnace, a Cornish engineer named Trevithick conceived the idea of placing the fire inside the large internal flue. He introduced this type which is known as the Cornish boiler.

The products of combustion pass from the fire on the grate bars C (Fig. 4) through the flue to the back end where they divide and return to the front end by means of the lateral flues L in the brickwork. See Fig. 4a. At the front the hot gases pass downward, and uniting pass through the flue F in contact with the bottom of the boiler. On leaving the boiler they go to the chimney. This arrangement of flues reduces the temperature of the gases before they come in contact with the bottom of the boiler where sediment collects. The grate bars rest on the dead plate D at one end

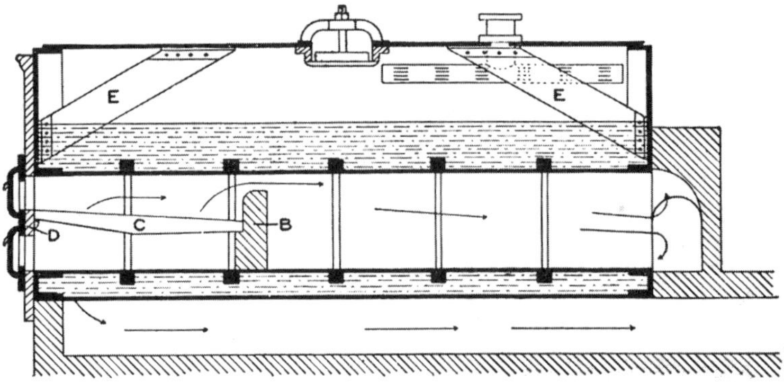

Fig. 4

and on the bridge B at the other; if made in two lengths (as is often the case) they are supported at the center by a cross bearer. The bridge is built of fire brick and the external flues are lined with fire brick. The heads are stayed to the shell by gusset stays E E.

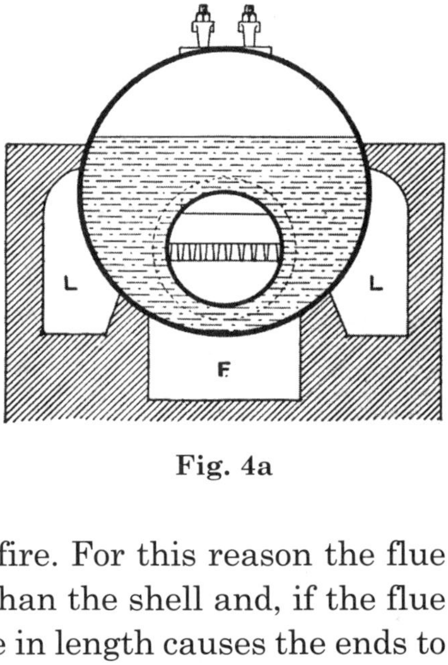

Fig. 4a

The large internal flue is the hottest portion of the boiler because it contains the fire. For this reason the flue has greater linear expansion than the shell and, if the flue is a plain cylinder, the increase in length causes the ends to bulge. When the boiler is cold, the flue returns to the normal length. This lengthening and shortening will soon loosen the flue at the ends. To overcome this, the flue is sometimes made up of several short rings flanged at the ends and joined by being riveted to a plain ring. This construction is shown in section in Fig. 4. Another method is shown in Fig. 5. The plain ring is riveted to the curved ring; this ring takes up the expansion, increases the heating surface, and strengthens

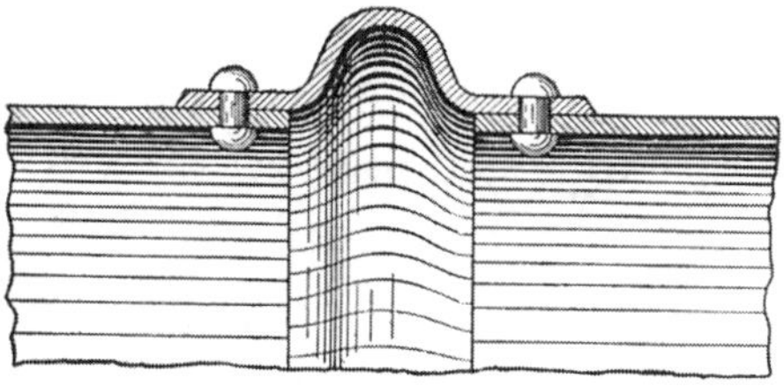

Fig. 5

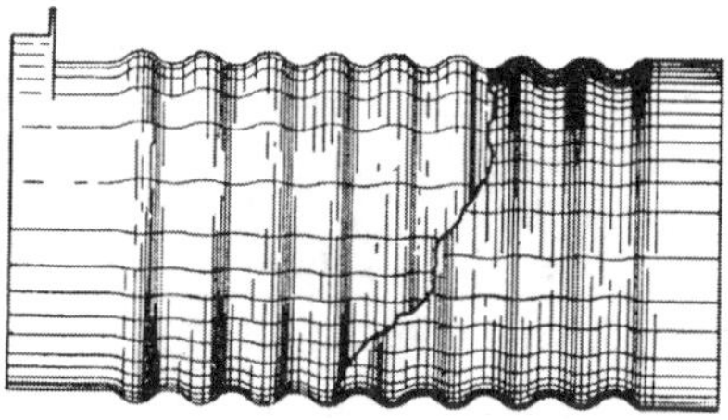

Fig. 6

the flue against external pressure. The same results may be obtained by the use of the corrugated flue, one form of which is shown in Fig. 6. The corrugated flue has many advantages over the devices shown in Figs. 4 and 5; it is frequently used in marine boilers.

Lancashire Boiler

Horizontal—Two-Flue—Internally-Fired

It can be proved, both by experiment and calculation, that with a given thickness large cylinders cannot stand as much external pressure as small ones. For this reason and on account of the short distance a fireman can throw coal accurately, the Cornish boiler is suitable for small powers only. If it is made too large, the flue is liable to collapse, but if, on the other hand, the flue is of too small a diameter, the grate will be insufficient. If this form of boiler is to be used in large size it is modified by using two flues instead of one. This boiler is called the Lancashire boiler. It is like the Cornish type except that it has two flues and, of course, two furnaces.

The flues are sometimes continued separately to the end. If they merge into one large flue, which forms the combustion chamber, it is called the "Breeches-flued" or duplex furnace boiler. These furnaces are fired alternately; the unburned gases set free from the freshly-fired coal are burned on meeting the hot gases from the incandescent coal of the other furnace. This arrangement prevents the escape of the unburned hydrocarbons.

The disadvantage of the Lancashire boiler is the difficulty in finding room for the two flues without greatly increasing the diameter of the boiler. Also, the small furnace is unfavorable to complete combustion as the space for the uniting and burning of the hydrocarbons is restricted. The combustion chamber of the breeches-flued boiler provides the necessary space, but the construction at the junction of the two flues is weak and has been the cause of many explosions.

Galloway Boiler

Horizontal—Two-Flue—Internally-Fired—Galloway Tubes

Another boiler of the same general form is the Galloway, shown in Fig. 7. This boiler differs from the Lancashire

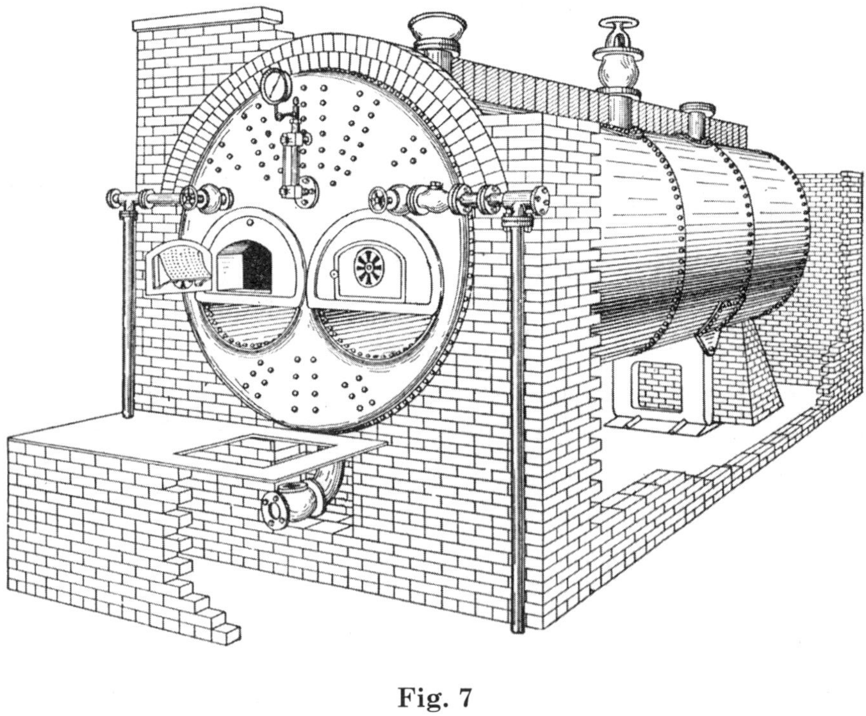

Fig. 7

in that short tubes are added to the flues. In the Galloway boiler having two distinct flues, the tubes were placed as shown in Fig. 8.

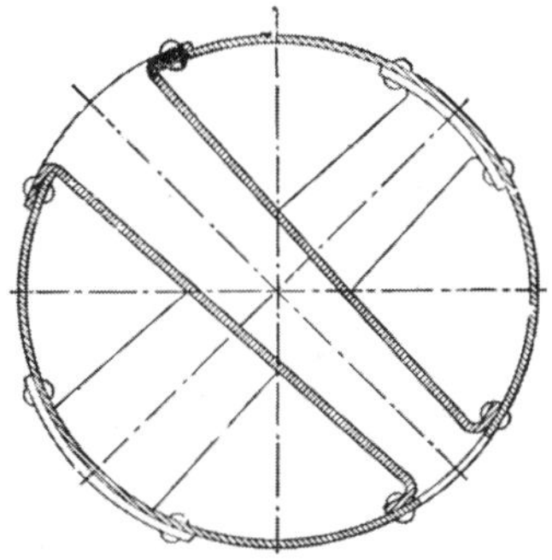

Fig. 8

In the later form of Galloway boiler, the two flues merge into one large flue of the shape shown in Fig. 9. This flue has corrugated sides and the conical tubes are staggered, thus insuring a thorough breaking up of the currents of hot gases. The tubes are made conical to facilitate removal for repairs. The shape of the tubes also permits the water to expand on being heated, and the particles rise vertically without disturbing the water on the heating surfaces above. The conical tubes are generally riveted rather than welded because the removal of a tube that is welded leaves a large hole in the flue.

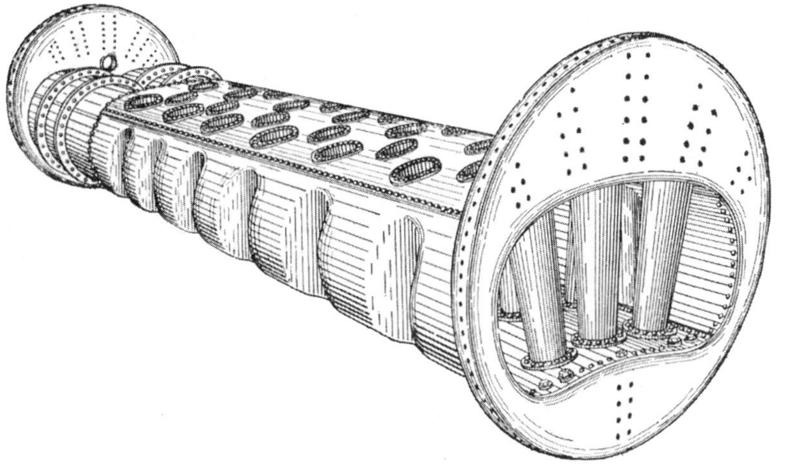

Fig. 9

Chapter 3

Fire-Tube Boilers

Single-Flue Boiler

Horizontal—Single Fire Tube—Externally-Fired

In the Cornish, Lancashire, and Galloway boiler the large internal flue served as a fire box. There was, however, a flue boiler having the fire external to the shell. The boiler shown in Fig. 10 resembles the plain cylindrical boiler both in appearance and setting, but it has one or more large flues extending from end to end. This flue increases the heating surface to such an extent that the boiler can be considerably shorter than the plain cylindrical.

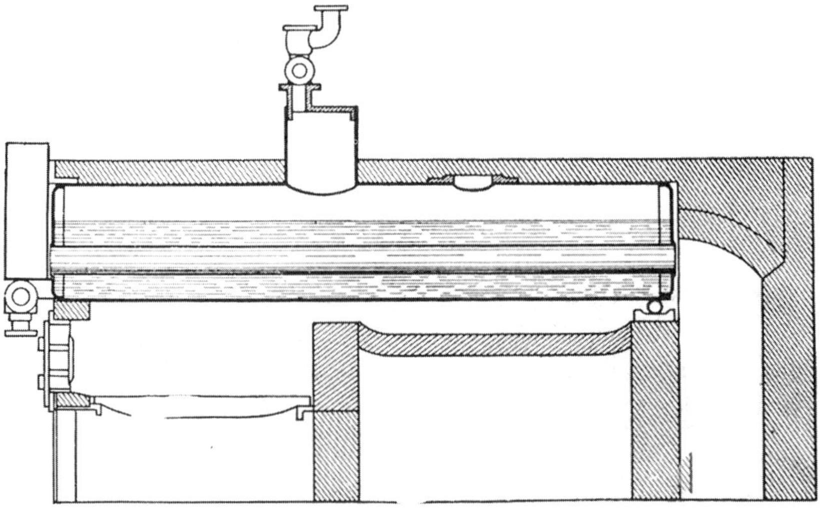

Fig. 10

Multitubular Boiler

Horizontal—Many Small Fire Tubes—Externally-Fired

When engineers found that the internal flue was such an advantage (that is, it increased the heating surface), they soon added more tubes; as the number increased, the size diminished until they became of the size used at present. This is in brief the development of the multitubular boiler. This type of boiler has many years been commonly used for stationary work and although other types possess advantages for certain conditions, it is still considered economical, reliable, easily handled, and safe if constructed of good material and operated with care and intelligence.

Figs. 11 to 14 are selected to illustrate this boiler. The boiler without the brick setting is shown in Fig. 11. It consists of a steel cylindrical shell and numerous small tubes extending from end to end. These tubes are 3 or 4 inches in diameter and are fastened to the two ends (called tube sheets) by expanding the tubes against the sheet and beading them over on the outside. The shell is made of steel plates 1/4 to 3/4-inch in thickness. At the front, the shell plates extend beyond the tube sheet and are cut away to allow the waste gases to enter the uptake. About one-third the volume of the boiler is occupied by the steam; the other two-thirds is

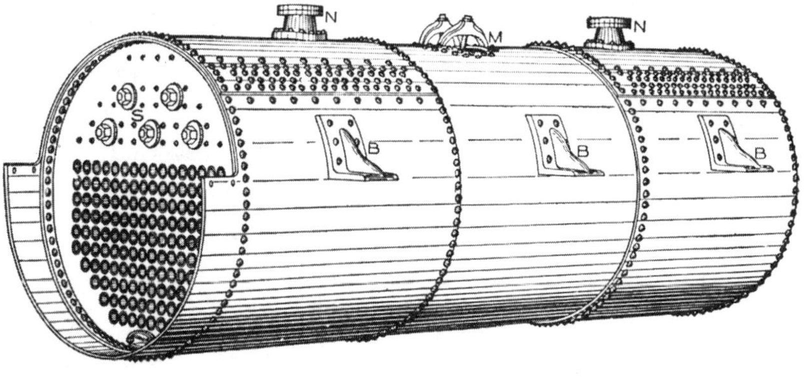

Fig. 11

filled with water and tubes. The water line is a little (from 4 to 8 inches) above the top row of tubes.

The flat ends are prevented from bulging by stays which may be of the form shown in Fig. 12 or they may be diagonal stays. The through stays are fastened to the tube plates by means of nuts and washers as shown at S in Fig. 11, and also in Fig. 12. Below the water level, the end plate is stayed by the tubes. This type of boiler may be supported by brackets B riveted to the shell or by means of beams and columns, as shown in Fig. 14. The front bracket is often fixed in the side wall, but the rear bracket should be placed on rollers so that it can move on an iron plate. This will prevent the straining of the plates from expansion and contraction. A small space must be left between the rear tube sheet and the brick wall to allow for expansion.

The boiler shown in Fig. 11 has two steam nozzles N. If the boiler has a dome (D Fig. 13) the steam nozzle is at or near the top of the dome. The feed pipe may enter either at the front or at the rear. It frequently terminates in a perforated pipe below the water line. The blow-off pipe is at the rear of the boiler as shown in Fig. 13. A valve, called the blow-off valve, regulates the flow and may be opened, when there is low pressure in the boiler, to blow out sediment and detached scale. The boiler is usually set with a slight inclination toward the rear so that mud and detached scale may collect near the blow-off pipe.

In order that the boiler may be entered for cleaning or repairs, it is provided with manholes and handholes. Fig. 11 shows a manhole M at the top near the middle and a handhole near the bottom of the front tube sheet. Handholes may be put in wherever desired, but manholes can be located only where the arrangement of stays and tubes will permit the entrance of a man. Manholes and handholes are

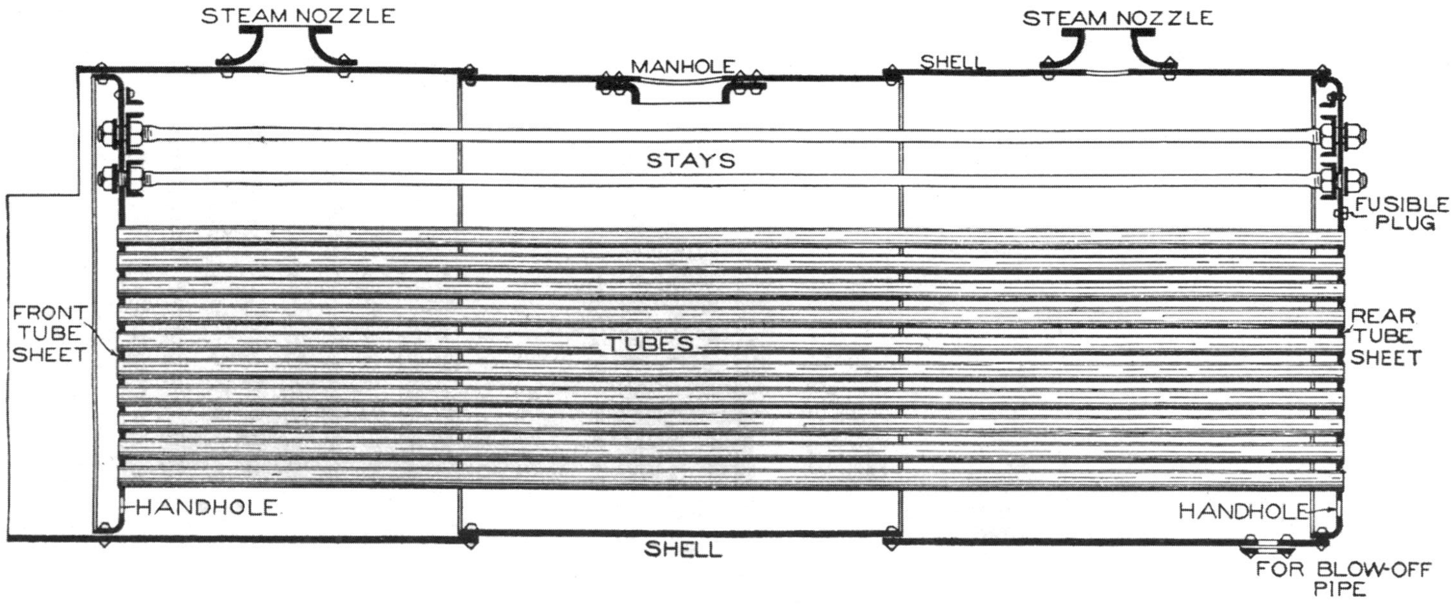
STEAM NOZZLE
MANHOLE
SHELL
STEAM NOZZLE
STAYS
FUSIBLE PLUG
FRONT TUBE SHEET
TUBES
REAR TUBE SHEET
HANDHOLE
SHELL
HANDHOLE
FOR BLOW-OFF PIPE

Fig. 12

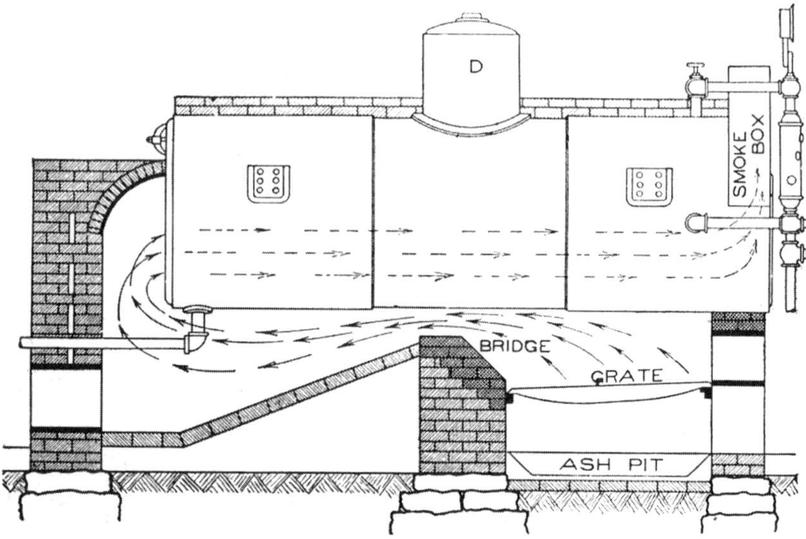

Fig. 13

elliptical; the former being about 11 inches by 15 inches in size; while the latter are about 4 inches by 6 inches.

The heating surface is the surface in contact with the hot gases. In this type, the heating surface is made up of about half the shell, the tubes, and about two-thirds of the rear tube sheet. In general, all the heating surface is below the water line.

The complete multitubular boiler is shown in its brick setting in Fig. 14, and a longitudinal section of the setting in Fig. 13. The brick setting consists of brick laid in cement or mortar. The bridge and the portions of the furnace exposed to the fire are lined with fire brick. The bridge is built at the rear of the grate and forms a support for the grate bars; it also directs the flames upward. The arrows show the direction of the flow of hot gases. The furnace is formed by the bridge, the side walls, and the lower part of the boiler front. The boiler front is usually of cast iron with the lower part lined with fire brick. The front has doors which lead to the furnace, ashpit, and smoke box. The space below the grate

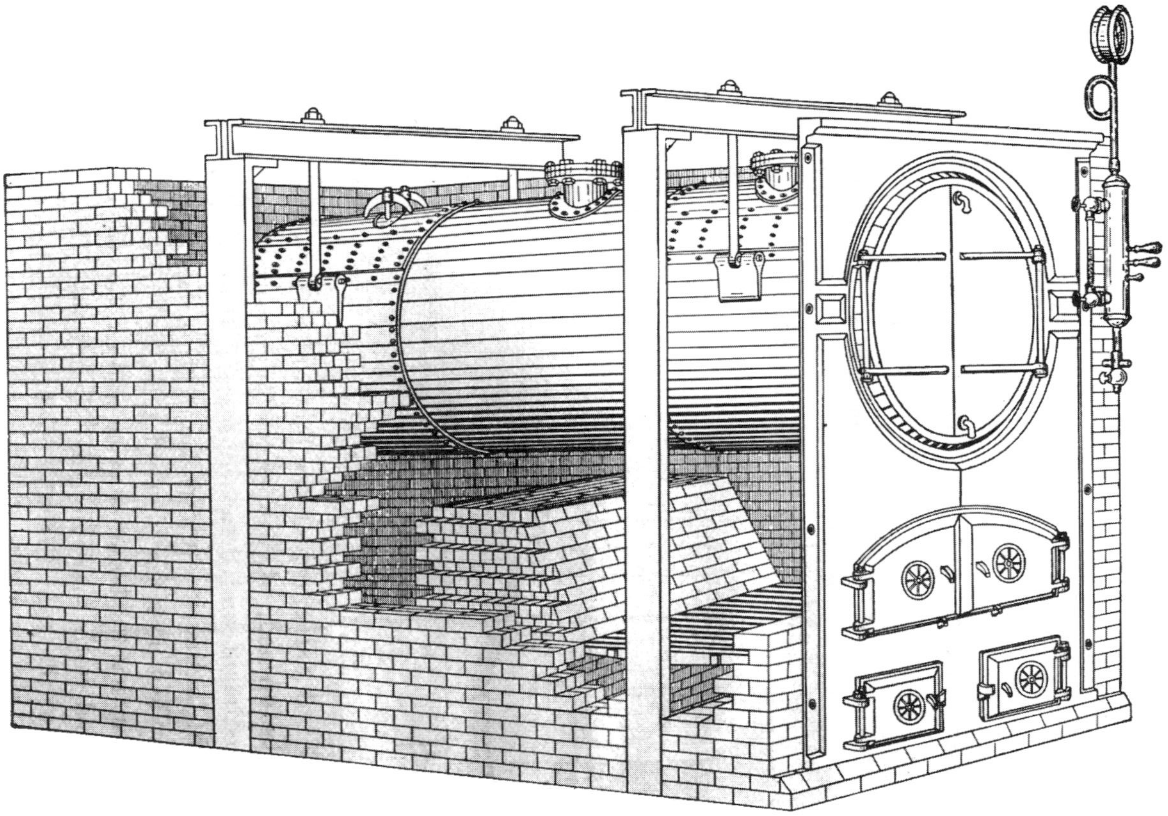

Fig. 14

is called the ashpit, and through its doors ashes are removed and a large portion of the air for combustion enters. Both the fire doors and ashpit doors have draft plates, or grids, to regulate the supply of air. The doors to the smoke box give access to the tubes for cleaning and repairs.

Upright Boilers

Vertical—Many Small Fire Tubes—Fire Box

Upright boilers are used when floor space is valuable and there is sufficient height. In small sizes, they are used for hoisting engines, pile driving, for supplying steam for pumps, and similar work; in large sizes when it is necessary to have a powerful battery in a small space. In general they are not as economical as the horizontal multitubular boiler unless they are carefully designed and of considerable height. If the tubes are short, the hot gases escape before they give up much of their heat.

One of the simplest forms of upright boiler is shown in Fig. 15. It has a cylindrical shell with a large fire box at its lower end. This fire box is formed by the inner cylinder which is fastened to the outer shell by short screw stay bolts as shown. A flanged ring connects the fire box with a large flue which conducts the hot gases away. The necessary handholes, gages, safety valves, etc., are provided. This form is not economical but is used on account of the little attention required.

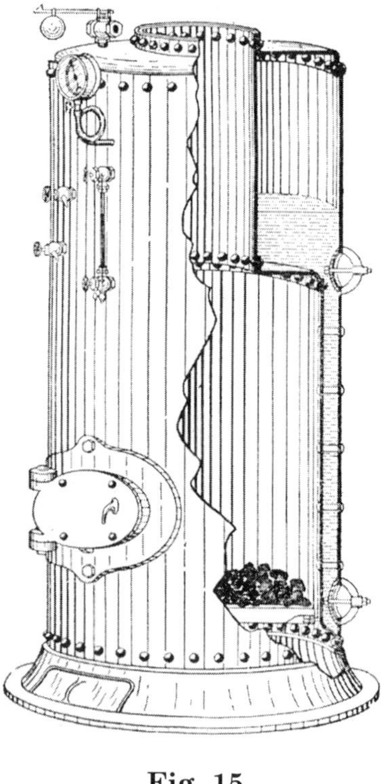

Fig. 15

More economical forms of the small upright boiler are illustrated in Figs. 16 and 17. The boiler shown in Fig. 16 is a common form; externally it is like the boiler represented by Fig. 15, but within, it has a somewhat different construction. It resembles a multitubular boiler placed on end. The fire box is made of an inner cylinder stayed to the outer. The top of the fire box, called the lower tube sheet, is connected to the upper head by tubes, through which the hot gases pass to the smoke pipe. It will be readily seen from Fig. 16, that the upper ends of the tubes are surrounded with steam while the lower portions are covered with water. As steam is a poor conductor of heat, the ends of these tubes are liable to injury from overheating.

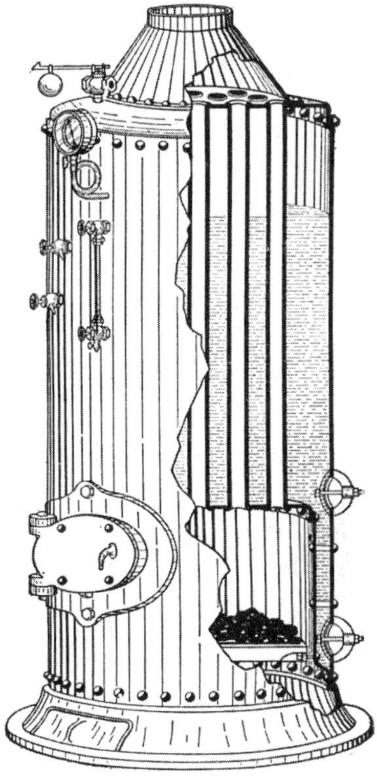

Fig. 16

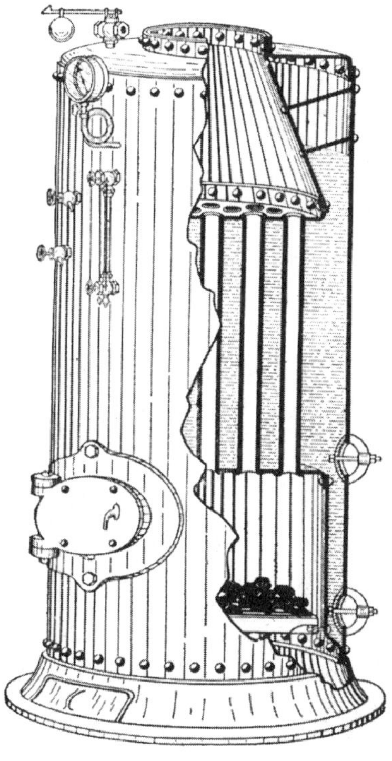

Fig. 17

In the class of boiler shown in Fig. 17 the upper ends of the tubes are below the water level, thus avoiding the weakness described in connection with Fig. 16. The upper tube sheet is submerged and is flanged and riveted to the frustum of the cone which forms the smoke box. The chief defect in this boiler is that the lower part of the cone is often placed too near the shell; this is done to admit more tubes. This construction restricts the space so much that there is not sufficient room for the steam to rise as it is formed on the tubes. The cone, which is subjected to external steam pressure, is likely to be weak and is usually carefully stayed.

These small upright boilers require no brick setting, as the fire box is within the boiler and the cast-iron foundation forms the ashpit.

Manning Boilers

Vertical—Many Small Fire Tubes—Fire Box

The Manning boiler is illustrated in Fig. 18. In order to get a large heating surface, it is made 20 to 30 feet high. It is, in general, similar to the upright boiler shown in Fig. 16. At the lower portion, the shell is of greater diameter than at the top in order to provide a large grate area. The inner fire box is stayed to the shell by screw stay bolts. As the fire box is surrounded by water and there are many long tubes there is a large heating surface. The tubes are arranged in concentric circles with a space for circulation in the middle.

The external fire box is joined to the shell by a double flanged ring as shown in Fig. 19; or, by the cone-shaped section as illustrated in Fig. 20. The top edge of the internal fire box is riveted to the lower tube sheet which is flanged. The bottom of the inner fire box is connected to the outer shell by a welded ring (shown in section in Figs. 19 and 20) called the foundation ring. The water space between the inner and outer fire box plates, called the water leg, should be large.

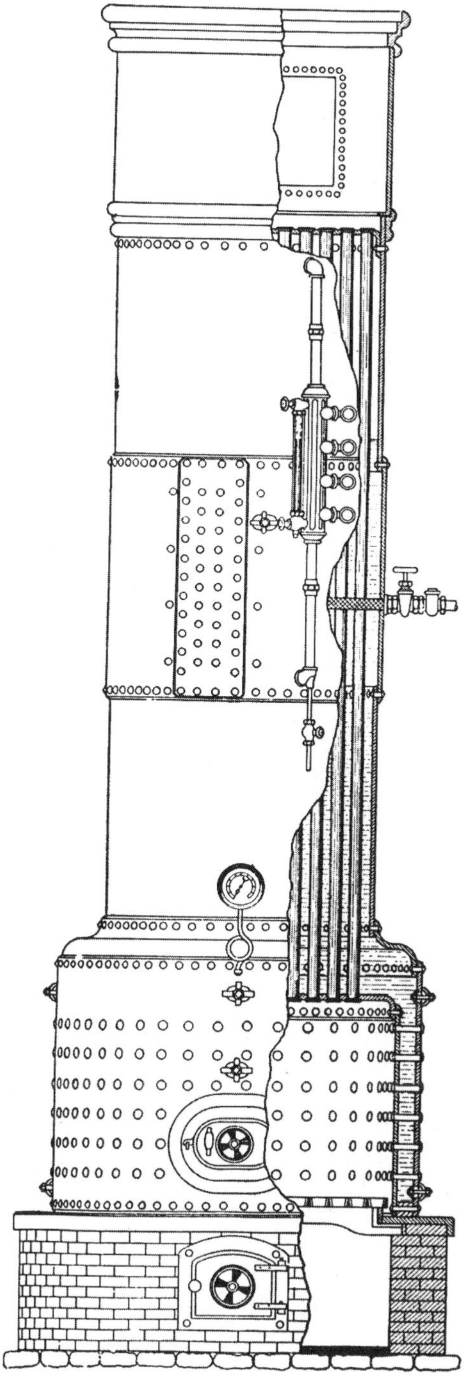

Fig. 18

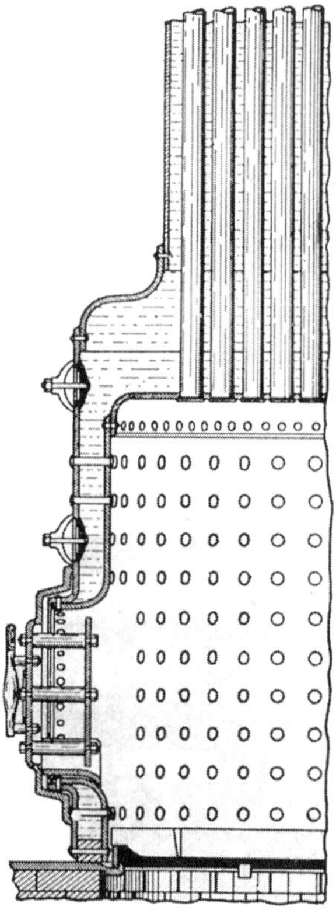

Fig. 19

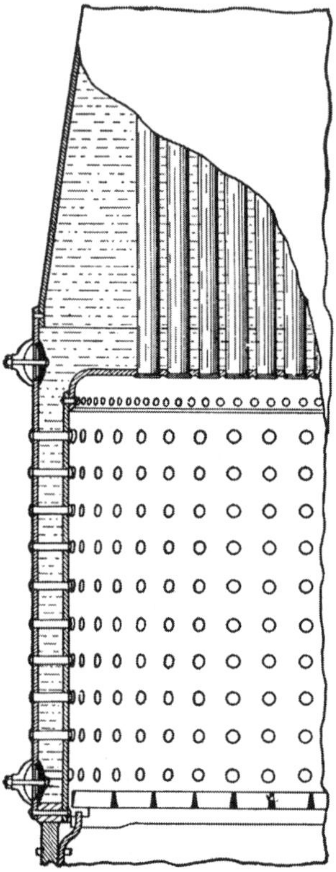

Fig. 20

This boiler is cleaned by means of handholes. They are placed in the shell plates near the lower tube sheet, in the external fire box just over the furnace door, and at the bottom near the foundation ring. As there are no manholes for cleaning, the boiler is suited to good feed water only.

The feed pipe enters the shell at the side near the middle of the water space, and extends across the boiler; it is perforated to distribute the water.

The heating surface consists of the inside of the fire box and the tubes up to the water level, and the tube sheet. That part of the tubes *above* the water line is the superheating surface; that is, the heat from the gases passes through the metal of the tubes to the steam, thus raising its temperature without raising its pressure. Steam heated under these conditions is called superheated steam. In small vertical boilers this superheating surface is not desirable because the work of the small boiler does not require superheated steam and the tubes are likely to be burned by the intense heat. With the long tubes of the Manning, the gases are not as hot when they reach the top, and as this boiler is built in large powers (200 horse-power being common) the engines supplied are built for economy and require dry if not superheated steam.

Return-Tube Boilers

Horizontal—Many Small Fire Tubes—Internally-Fired

The boilers hitherto described are used mainly for stationary work, the exceptions being so few that they need not be even mentioned. Let us now discuss another modification of the fire-tube boiler—one that has been and is now extensively used in marine work. The parts of the return-tube boiler are essentially the same as those of flue boilers (Cornish and Lancashire) and the multitubular boiler. They are, however, arranged differently in order to be used on board ship.

The earliest forms of marine boilers, working with pressures of 15 to 30 pounds per square inch, were square or box-shaped. They were economical and of convenient form for ships. When higher steam pressures became necessary, the flat surfaces required so much staying that they were abandoned and the cylindrical type introduced, as this form is the best of the practical shapes to resist internal pressure. The cylindrical form may not be as conveniently stowed aboard ship, but it will stand much higher pressures. The cylindrical marine boiler is frequently built for 170 pounds per square inch.

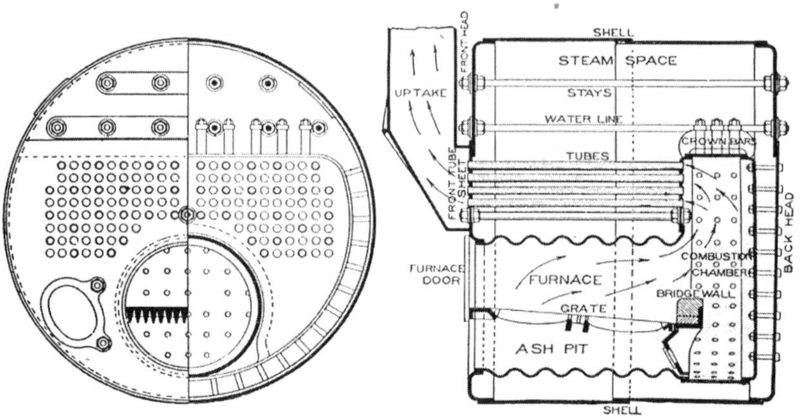

Fig. 21

The single-ended, return-tube boiler, shown in Fig. 21, combines the internal furnace flue of the Cornish type and the numerous small fire tubes of the multitubular. The cylindrical shell is made up of plates riveted together and to the flat ends of the boiler, which are flanged to fit the shell.

The furnace is cylindrical, three to four feet in diameter and about seven feet in length. The front end of the furnace flue is riveted to the front end plate, which is flanged for the purpose. The back end is riveted to the combustion chamber plates. Formerly, the flue was a plain cylinder, but as a plain cylinder, unless of small diameter, cannot stand much external pressure, it soon became necessary to strengthen it. This was done by means of the curved ring shown in Fig. 5 and other methods; but at present the corrugated flue is used, one form being shown in Fig. 6.

The grate is placed at about the center of the height of the furnace flue; the space above this grate is occupied by the fire and hot gases, below it is the ashpit. As will be seen from the arrows in Fig. 21, the hot gases fill the space above the fire, the combustion chamber, the tubes and the uptake.

The combustion chamber in which the products of combustion are completely burned, is formed of flat and curved plates flanged at the edges and riveted together. The shape of the plates is shown in Fig. 21, which is a sectional view of a single-ended marine boiler. The back tube sheet forms the front of the combustion chamber. The space around the tubes, furnace flue, and combustion chamber is filled with water, the water level being six to eight inches above the top row of tubes. The space above the water level is called the steam space.

As the return-tube boiler has several flat surfaces, this type requires careful staying. The flat ends above the water level are prevented from bulging by long stay rods which are similar to those in the multitubular type. Below the water level, the furnace flue and the tubes aid in holding the flat plates together. In addition, a few of the tubes (shown by the heavier circles in Fig. 21) are made thicker so that a thread may be cut on the ends which are screwed into the tube sheets and held by thin nuts. The combustion chamber plates are stayed to the rear end plate and the shell by short screw stay bolts. The flat top of the combustion chamber is supported by girders or crown bars.

Number of Furnaces. The boiler shown in Fig. 21 has only one furnace, but return-tube boilers frequently have two, three, or four furnaces.

Fig. 22 shows a boiler with three furnaces. Large furnaces are more efficient than small ones because the grate area increases directly as the diameter, while the air space above the grate increases as the square of the diameter. The greater space aids combustion. The length of the grate bars is nearly constant for all sizes of flue because it is limited by the distance a fireman can throw coal. Furnace flues are usually from 36 to 54 inches in diameter. As the size of furnaces is fixed, the number depends upon the size of the

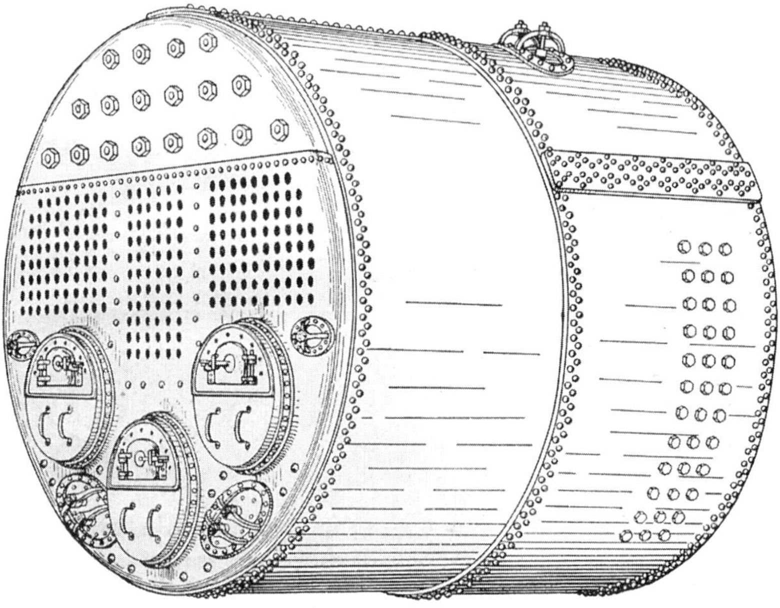

Fig. 22

boiler, for a large boiler must have a large grate area which can be obtained only by using several furnaces. The various arrangements are shown diagrammatically in Fig. 23.

A single-furnace boiler has but one combustion chamber. A two-furnace boiler may have a combustion chamber for each furnace or it may have a common combustion chamber. If there is but one boiler on board, it is better to have two combustion chambers, so that in case a tube bursts, the boiler will not be disabled. If, however, there are several

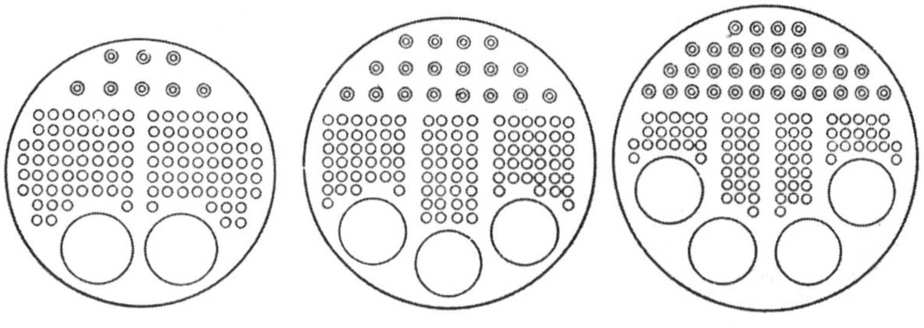

Fig. 23

boilers, it is better to have a common combustion chamber for the two furnaces, because the alternate stoking keeps up a more nearly constant pressure of steam and there is less smoke. Three-furnace boilers usually have three combustion chambers, while four-furnace boilers have two. In case four furnaces are used with three combustion chambers, the two center furnaces lead to a common combustion chamber and each outside furnace has one.

Double-ended Boilers. This form of marine return-tube boiler is practically the same as two single-ended boilers placed back to back, but with the rear plates removed. The weight of the rear plates is saved and there is less loss from radiation. This makes the double-ended boiler lighter and cheaper in proportion to the heating surface. Double-ended boilers are often made 16 feet in diameter and 18 feet long.

There are two distinct classes of double-ended return-tube boiler—those having all the furnaces open into one combustion chamber and those having several combustion chambers. The boiler having but one combustion chamber has the disadvantage that if one fire is being cleaned the whole boiler may be cooled by the inrush of cold air. It is better to have a combustion chamber for each furnace or at least have a combustion chamber for the furnaces of each end. The usual method of dividing up the combustion chambers is by water spaces as shown in Fig. 24, which is the section of a boiler having a combustion chamber for each furnace.

Internal Furnace Return-Tube Boiler. Although the return-tube boiler is commonly used in marine work, this type, with some changes in detail, is used in plants ashore. Fig. 25 shows the construction and arrangement of parts. The flue is larger in proportion to the diameter than is the case with the marine form; the combustion chamber

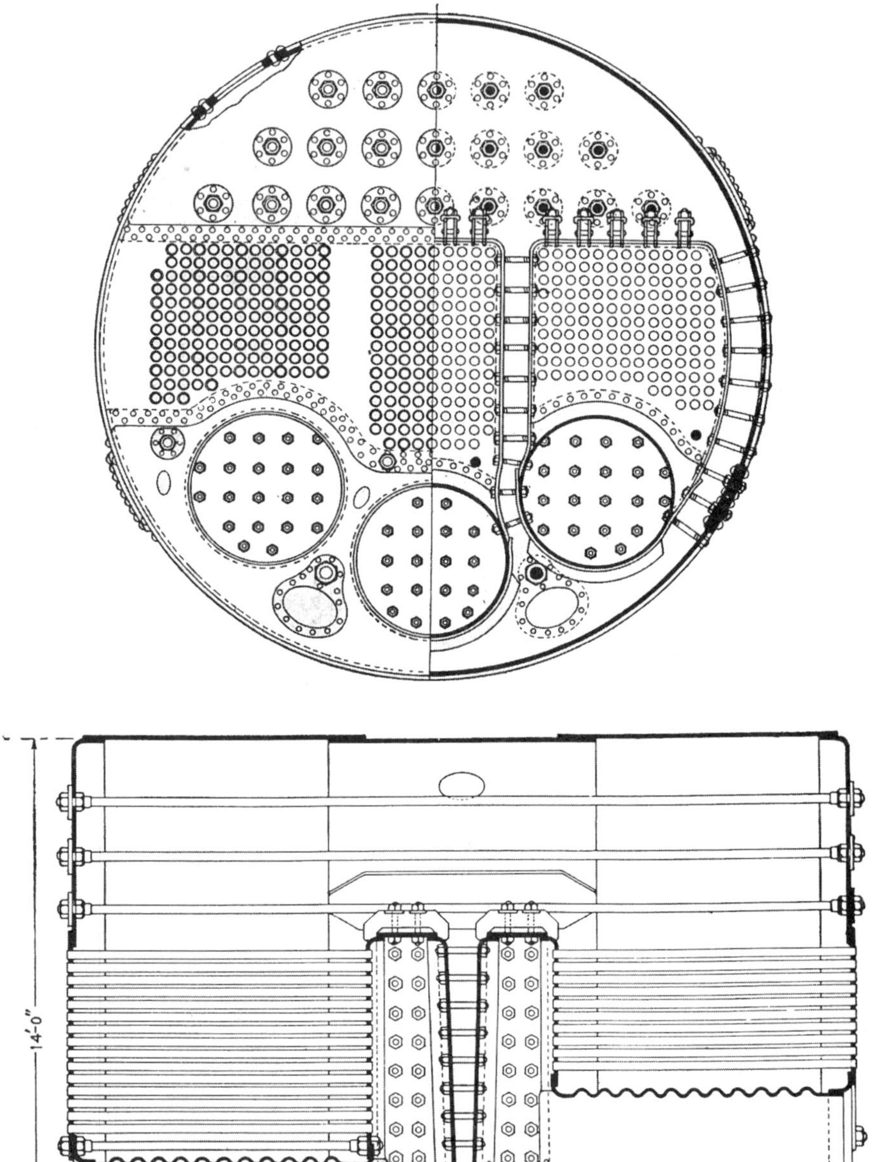

Fig. 24

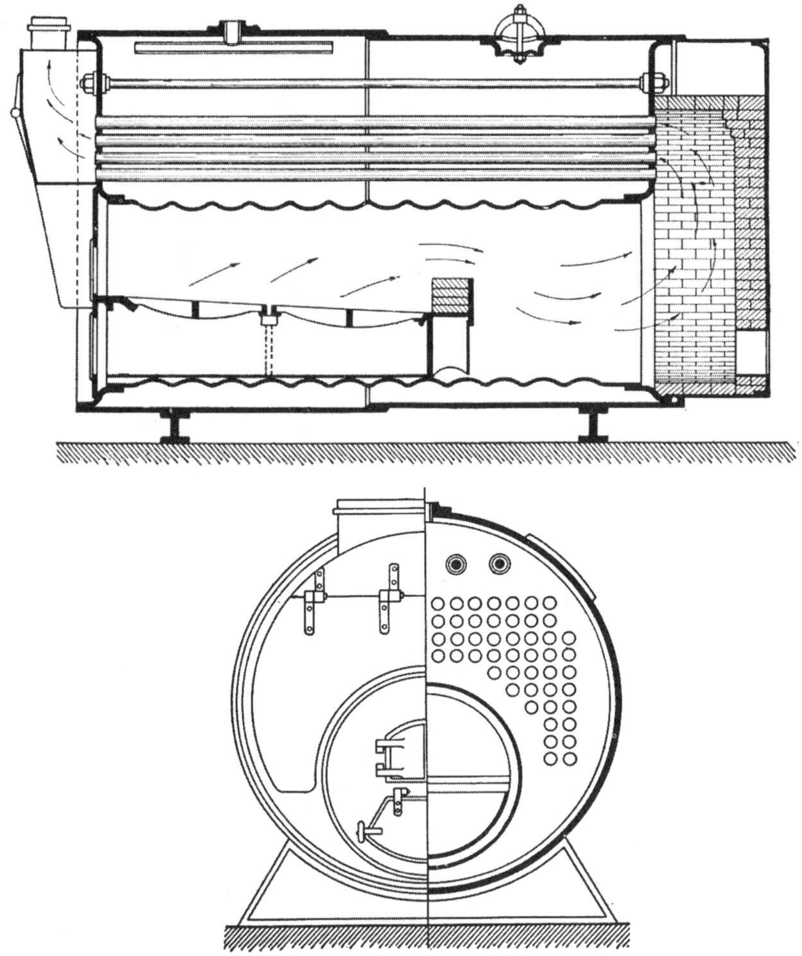

Fig. 25

is partly external to the shell, that is, the rear tube sheet is also the rear end plate. This arrangement does away with the necessity of staying the flat plates of the combustion chamber.

Another form of internal furnace, return-tube boiler is shown in Fig. 26. This boiler usually has two flues extending from the front to the back head. The grate is placed in the corrugated portion while conical water tubes support the flue back of the bridge wall. The large furnaces and the

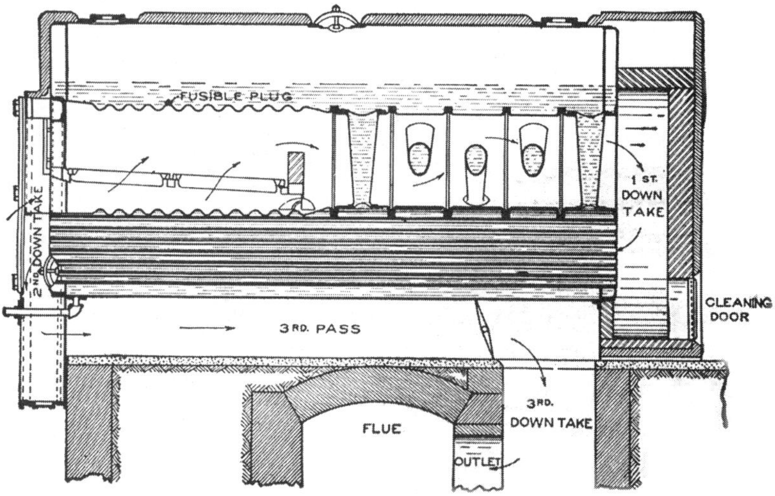

Fig. 26

space around the conical tubes provide a combustion chamber of ample size.

The arrows show the direction of the hot gases. After leaving the internal flue they enter the return tubes which are below the furnace; before leaving the boiler, they pass underneath the shell. By this arrangement the hottest gases are near the water line and the cooler gases in contact with the cold water, thus there is the greatest difference in temperature at all times. At each change in the direction of the hot gases, there is an opportunity for dirt and ash to fall by gravity so that the tubes may remain clean and efficient.

With the exception of the foundation there is no brickwork. The shell is covered with a non-conducting material. This boiler, like the Galloway, has a large steam and water space, thus insuring dry steam and great reserve power.

Through-Tube Boilers

Horizontal—Many Small Fire Tubes—Internally-Fired

Vessels of slight draft require a boiler of small diameter. This is especially true of gunboats as it is desirable to have the boilers below the water line. As there is not room

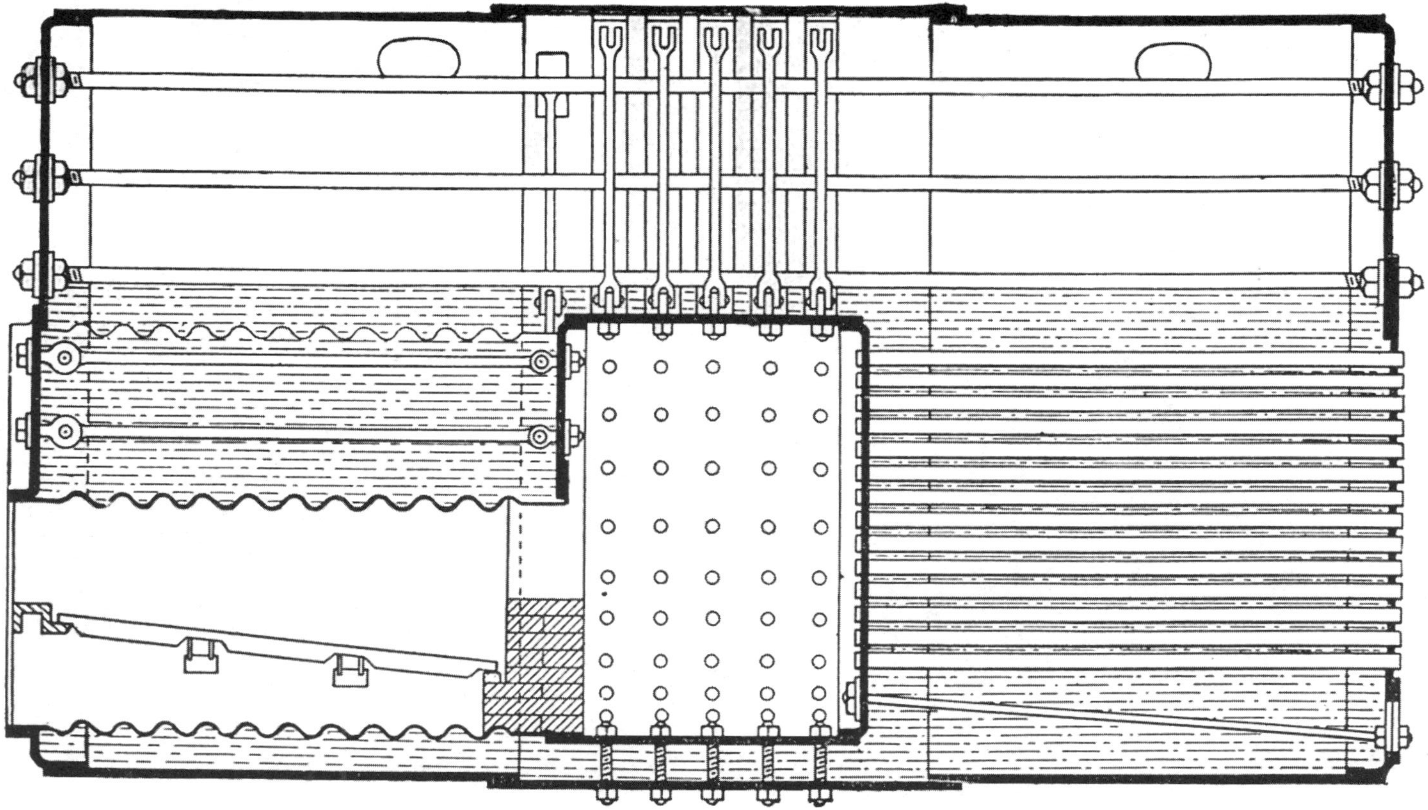

Fig. 27

for the return-tube boiler, the through-tube, shown in Fig. 27, is sometimes used. This boiler is made up of the same parts as the return tube, the chief difference being that of arrangement. The rear plate of the combustion chamber forms one tube sheet and the end plate forms the other. The top of the combustion chamber is stayed to the shell by sling stays which are bars having forked ends fastened to the shell and to the combustion chamber.

The fire is in a flue, or flues, which leads to the combustion chamber. The hot gases pass from the combustion chamber through the tubes to the uptake at the back end. The chief objection to this form is its length, for the heating surface is small unless the boiler is made very long.

Fire-Box Boilers

Locomotive Type

Horizontal—Many Small Fire Tubes—Externally-Fired

Although vertical fire-tube boilers may be classed as fire-box boilers, the name fire-box boiler is usually applied to the locomotive type whether used with a locomotive or as a stationary boiler.

The usual form of horizontal fire-box boiler consists of a cylindrical shell, or barrel, partly filled with tubes, and a rectangular fire box. The shell is prolonged beyond the rear tube sheet to form a smoke box. The front ends of the tubes open into the fire-box, while the rear ends open into the smoke box. The hot gases from the fire pass through the tubes to the smoke box and from thence to the stack or uptake. For locomotive work, there are a large number of small tubes (usually 2-inch), but for stationary work the tubes are larger and less numerous. The reason for this difference is that in the locomotive boiler a greater heating surface is necessary, and to obtain sufficient draft to burn the large amount of coal for this heating surface, the exhaust steam is turned into the smoke box. The blast of steam carries the heated gases up the stack and a fresh supply of air passes through the grate.

The cylindrical shell is joined to the fire box by riveting to a flanged ring or to a cone-shaped portion as in the vertical boiler. The fire box has a rectangular cross-section and usually a flat top. Like the vertical boiler there is an inner and an outer fire box, the inner having the same shape as the outer, except that the top is flat. The external fire box is connected to the inner by short screw stays. The space between is called the water leg. The flat top is stayed by girders or crown stays. These are sometimes attached to the shell by sling stays. The lower portions of the tube sheets are held in place by the tubes; the upper portions are stayed by diagonal stays.

The chief differences in the various forms of locomotive boilers are the shape of the fire box and the location of the grate. Locomotive boilers are either straight top or wagon top. The wagon top boiler, see Fig. 28, has a cone-shaped portion by means of which the boiler is larger at the firebox end. This construction is to give a greater steam space. The increase in size of boilers has raised the top so high above the rails that the wagon top is not now used extensively; the straight top, see Fig. 29, is more common.

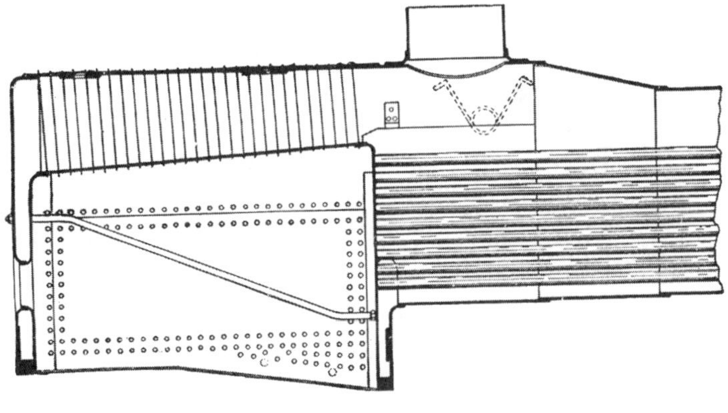

Fig. 28

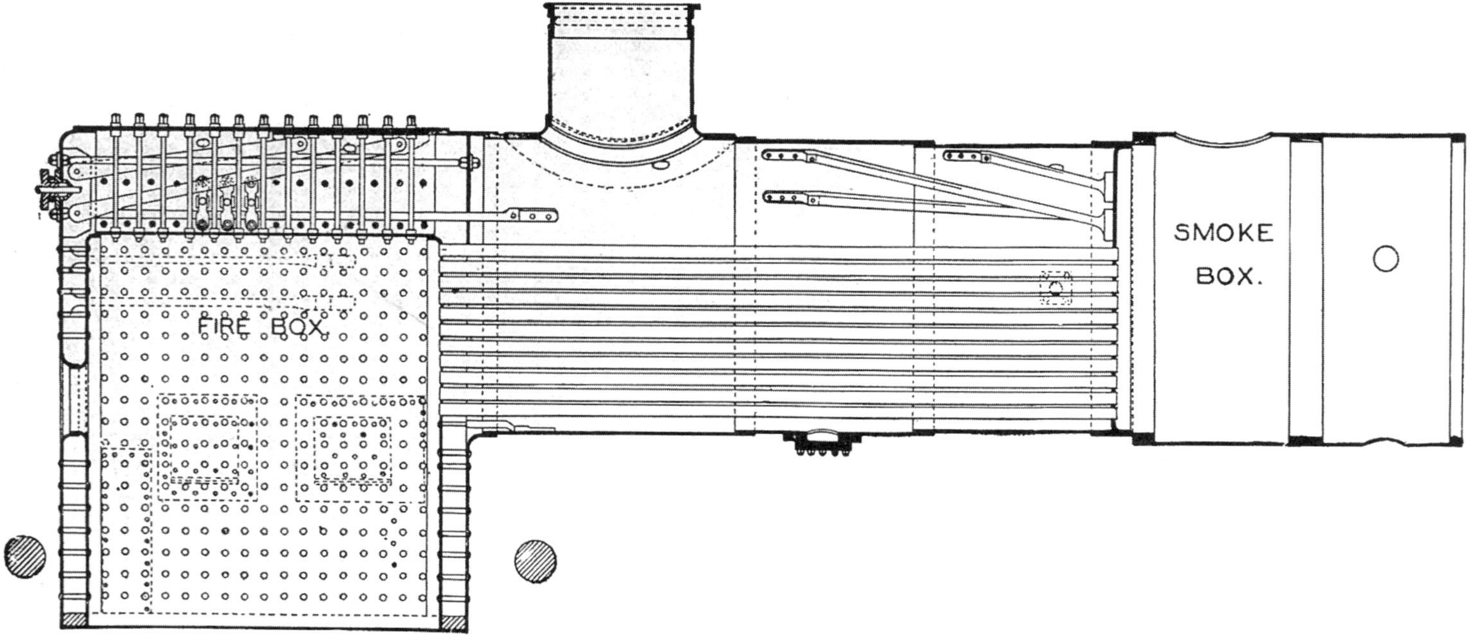

Fig. 29

Belpaire Boiler. The shell and fire tubes of this type of boiler are practically the same as in any other fire-box boiler, the peculiarity lies in the fire box. The inner and outer fire-box plates are horizontal at the top, and the sides of the outer fire box are continued so that the space above the crown sheet is rectangular in section. The advantage of this construction is that the staybolts holding the crown sheets and side sheets can be placed at right angles to the sheets. This reduces the tendency to bending when under pressure.

Wootten Boiler. In this type also, the fire box is the chief portion to be considered. The size of the locomotive fire box is limited. With older types the width was limited to less than three feet and the length to less than seven feet. This was because of the frames and the distance between the axles. By placing the fire box above the axles, the width was increased by an amount equal to the thickness of the frames or about seven inches, and the length increased to about eleven feet. By raising the fire box still more and placing it above the driving wheels, the width can be still further increased.

A broad, shallow fire box is required if anthracite coal is used. The Wootten fire box, shown in Figs. 30 and 31, is very wide and is placed on top of the driving wheels. Formerly, a combustion chamber was placed between the end of the grate and the tubes, but as it was found to be unnecessary, it is not now used.

Lentz Boiler. The object of the design shown in Fig. 32 is to avoid the use of stays. To do this no flat plates are used, except the tube sheets and these are stayed by the tubes. The fire box is in the form of a corrugated flue similar to those in internally-fired, return-tube boilers. As this is circular it requires no stays. The shell is circular and shaped as shown in the illustration. This type has been much used in Europe, but a few have been built in this country.

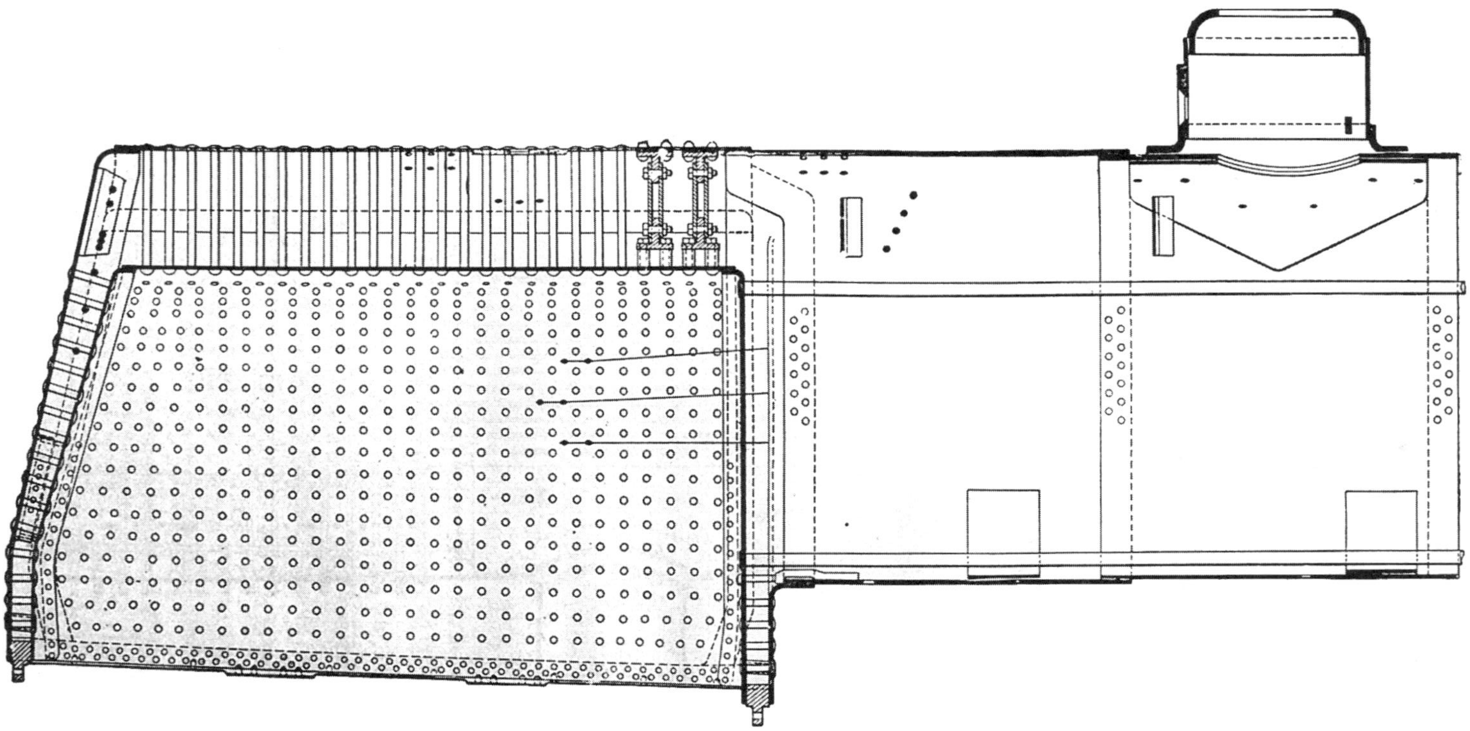

Fig. 30

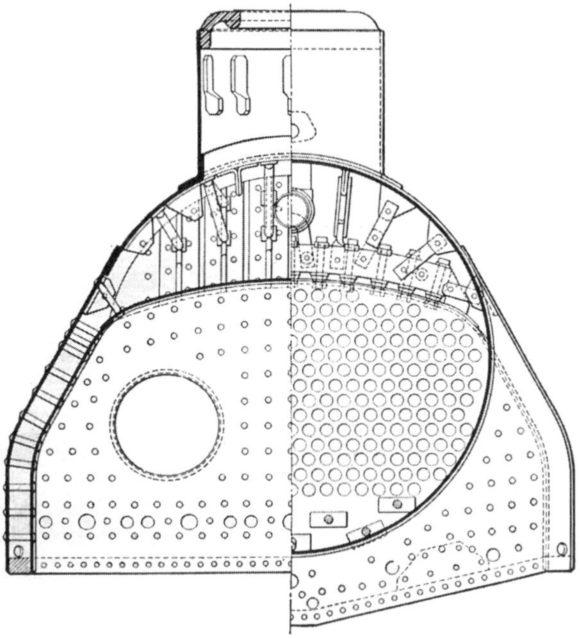

Fig. 31

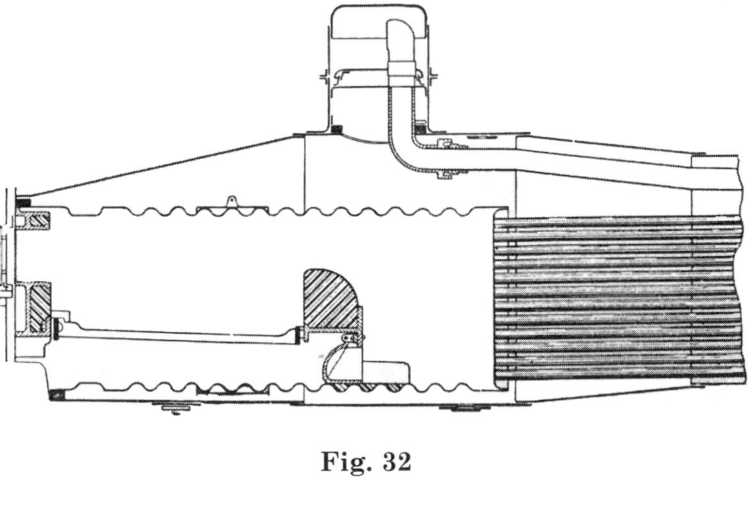

Fig. 32

Fire-box Boiler for Stationary Work

The fire-box boiler, usually called the locomotive boiler, is often used for stationary work, traction engines, and for vessels of light draft. This type of boiler, slightly modified, is sometimes used for generating the steam for heating buildings. It is economical and durable when used with natural draft. The chief differences in construction are larger tubes because of the draft, and the changes due to method of support. A common form is shown in Fig. 33. This type has been built in large sizes for high pressure, but when so made is expensive.

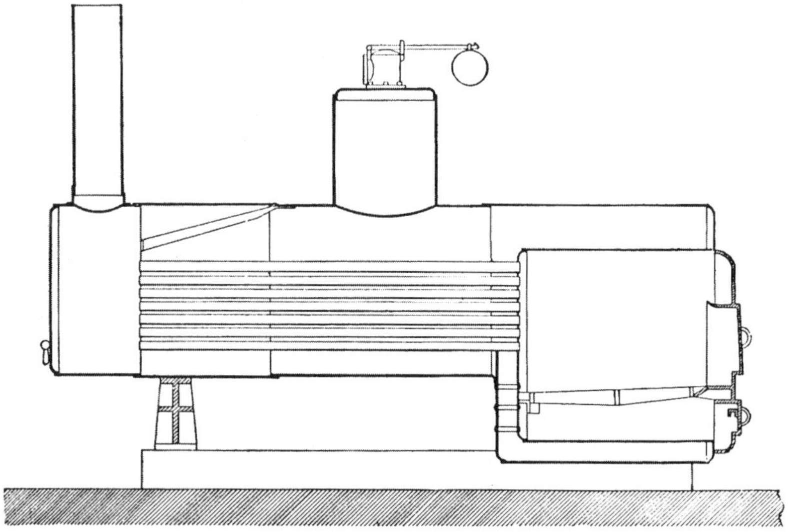

Fig. 33

Peculiar Forms

Fire-Tube Boilers, but differing from those described

Return Tubular as Stationary. Boilers of the form shown in Fig. 34 resemble the locomotive, fire-box type, but in addition have return tubes. The hot gases reach the uptake by means of these tubes instead of passing to the chimney from the smoke-box end. Thus they combine the advantages of the fire-box type and the return-tube type

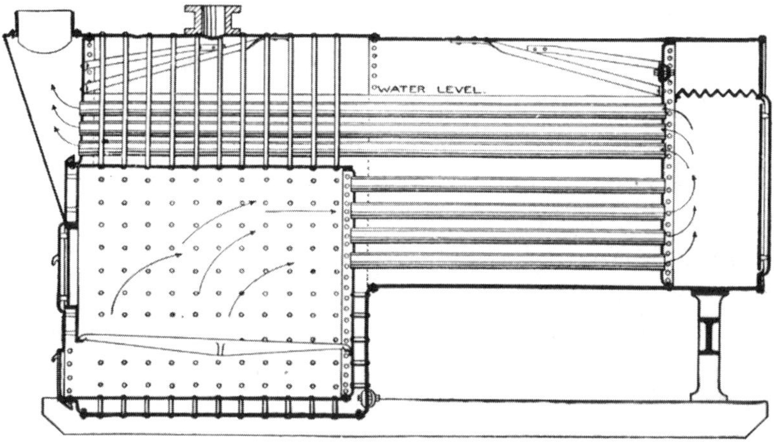

Fig. 34

without the brick furnace. The water surrounds the furnace on all sides except the front. They are built in sizes from 12 to 70 horse-power. As Fig. 34 shows the construction so clearly, further description is unnecessary.

The Cochrane Vertical Boiler is somewhat like the return-tube boiler in point of arrangement of heating surface. This boiler is shown in section in Fig. 35. The hot gases pass from the furnace to the combustion chamber, then through the tubes to the uptake. The heating surface consists of tubes and the plates of the fire box which is surrounded by water except the bottom. The crown of the boiler and of the fire box, being hemispherical, require no staying. The hemispherical crown also allows a large steam space. The flat plates (the tube plates) are held together by the tubes.

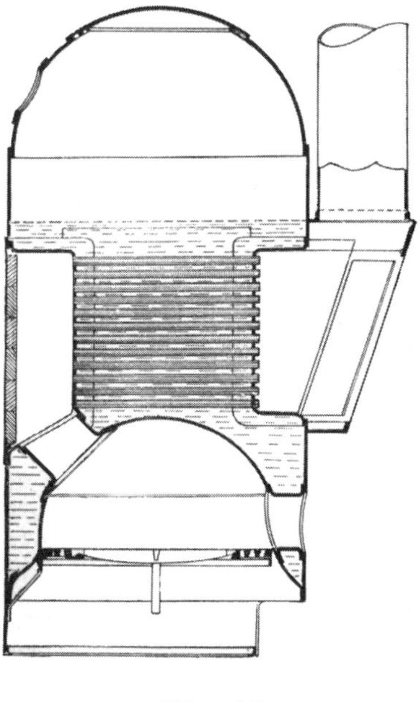

Fig. 35

The Shapley Boiler, shown in Fig. 36, may be called a return-flue vertical boiler. The upper portion is a reservoir for water and steam and the lower contains the fire box. The crown sheet of the fire box is stayed to the top by through stays. The hot gases from the fire rise in the fire box, pass through short horizontal tubes to an annular space. This annular space is connected to the flue at the base by vertical tubes passing through the water space.

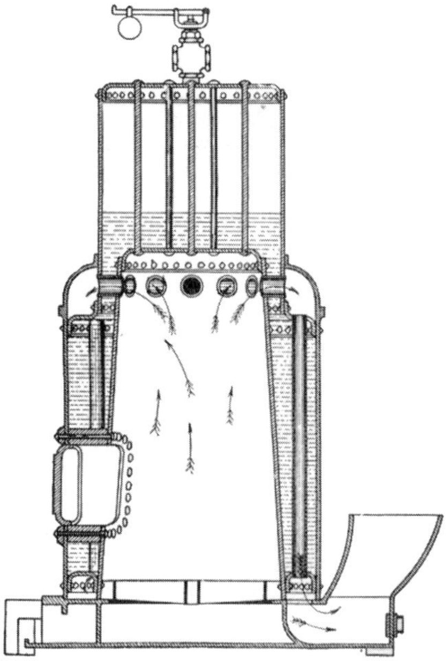

Fig. 36

The boiler has a large combustion chamber; the fire box is surrounded by water, and the crown sheet and the tubes are removed from the intense heat of the fire. This arrangement increases the heating surface, allows complete combustion and results in a durable boiler. The base is partially filled with water so that any sparks carried over will be quenched.

Robb-Mumford. This boiler resembles the through-tube internally-fired boiler (see Fig. 27) in that the fire is within a corrugated flue and the tubes lead from this flue to the rear of the boiler.

The Robb-Mumford boiler consists of two cylindrical drums or shells, connected at each end by a neck. See Fig. 37. The upper drum is a steam and water drum, the water level being at about the middle. The lower shell is larger and is inclined about one inch in twelve to increase the circulation, and facilitate washing out. In this lower shell is the

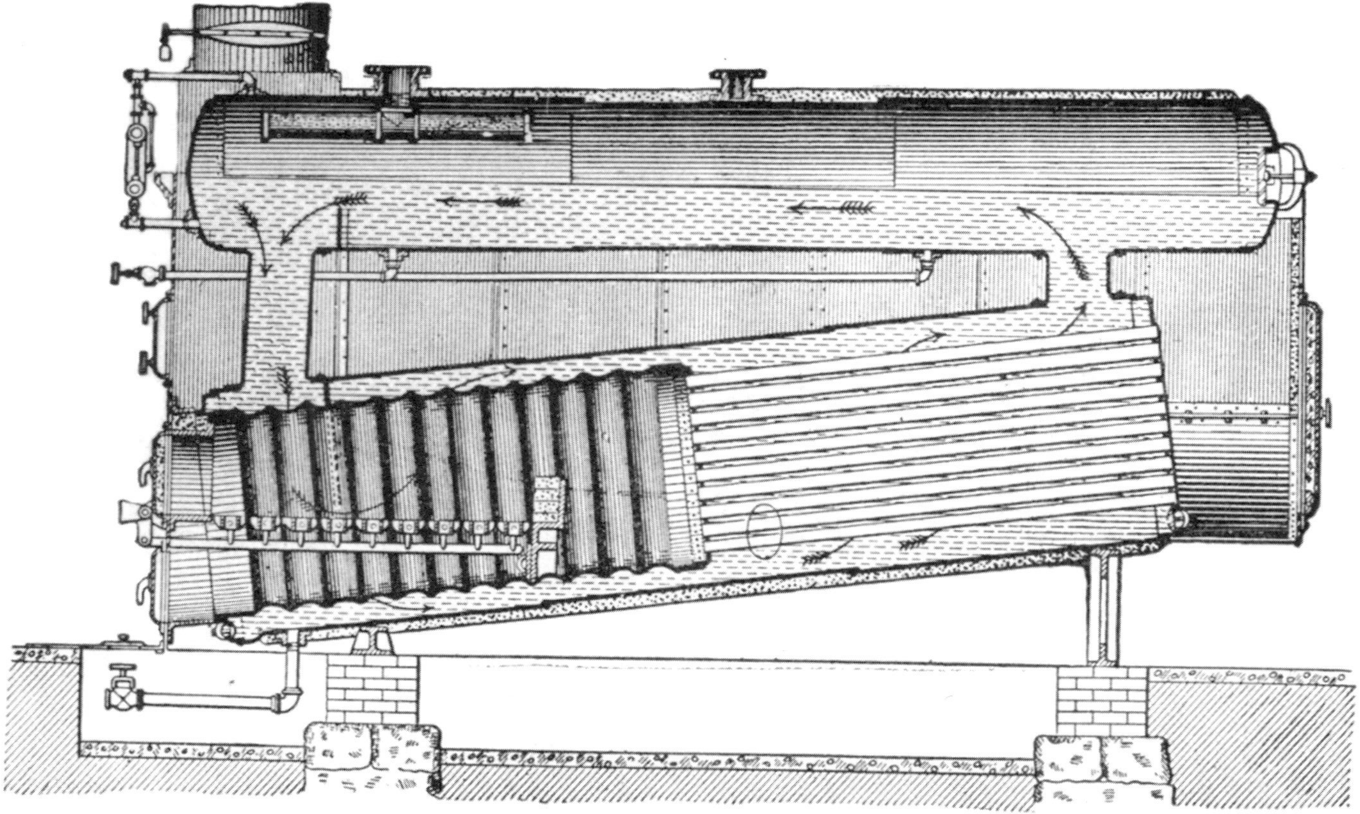

Fig. 37

corrugated flue containing the grate. The fire tubes nearly fill the remaining portion of the shell as shown.

The furnace, in which the coal is burned, is surrounded by water; the hot gases pass through the tubes to the rear, return between the lower and upper shells and escape to the chimney from the front to the boiler. The steel casing keeps the gases in contact with the drums. The water circulation is shown by the arrows. The mixture of water and steam enters the upper drum; the steam here separates from the water which flows down the neck at the forward end. A semicircular baffle plate directs this water around the furnace to the bottom. The feed water enters at the rear of the steam and water drum.

As compared with the multitubular boiler one can readily see that the drums can be made much thinner on account of the small diameter. The tubes are short, straight, and easily cleaned. The internal furnace does away with much of the loss from leakage. As the water is well subdivided, steam can be raised rapidly and there is little danger of a disastrous explosion.

Directurn. The Begg's "Directurn" boiler (Fig. 38) is, in brief, a horizontal, externally-fired, multitubular boiler in which tubes conduct hot gases through the space behind the bridge. This boiler consists of a shell partly filled with 3-inch tubes. The rear of the furnace is a throat sheet in which 4-inch tubes are expanded. The other ends are expanded into the rear end plate which is made large enough for the purpose. The boiler is encased in steel plates lined with fire brick which is held in place by rods passing through the notches as shown. The manhole for entering the boiler is placed in the front head instead of in the shell as is frequently done.

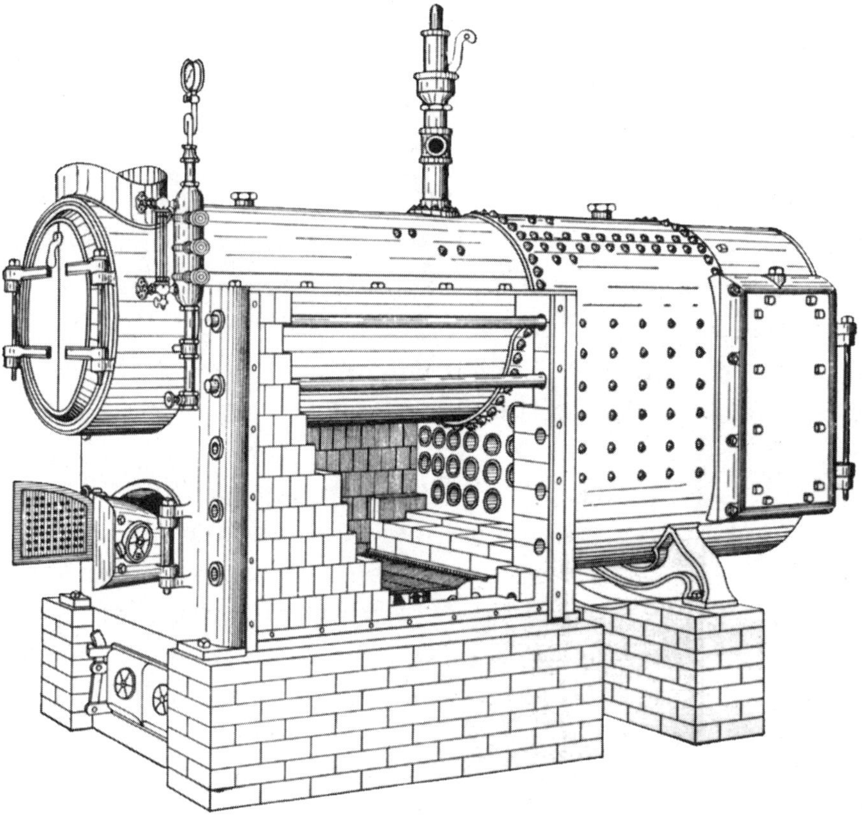

Fig. 38

Chapter 4

Water-Tube Boilers

The water-tube boiler differs essentially from the fire-tube. The names indicate the chief point of difference. In the fire-tube boiler, the tubes, which are surrounded with water, conduct the hot gases to the smoke box. In the water-tube, the tubes are filled with water, and the hot gases pass over and among them on their way to the chimney.

Although flue boilers and the tubular types were introduced at an earlier period than the water-tube, the last-named type is not a new form of steam generator. About a century ago, John Stevens invented a water-tube boiler and fitted it to a steamboat. This boiler (Fig. 39) was a combination of small tubes connected, at one end, to a reservoir. Thus the "porcupine" was one of the earliest forms. At various times since then, many ideas have been worked out both for marine and stationary boiler. During the last

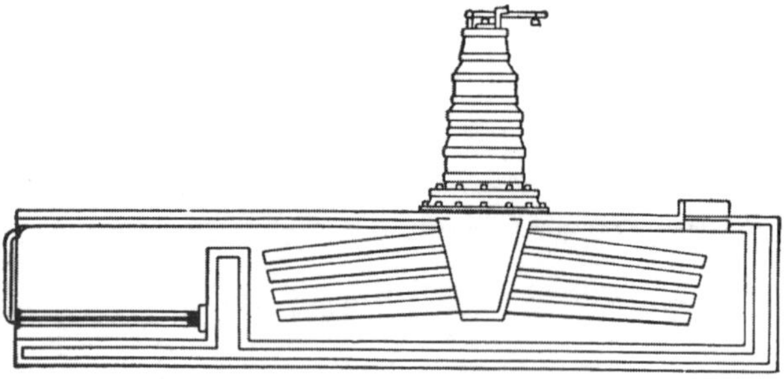

Stevens Boiler

Fig. 39

fifteen years, however, the water-tube boiler has been steadily growing in favor, the chief reasons being—the necessity of higher steam pressures, greater reliability of materials, greater skill in design and workmanship, and more intelligent management.

It is not within the province of this instruction paper to discuss the relative merits of fire-tube and water-tube boilers, but a careful, impartial consideration seems to show that as far as economy of running is concerned there is but little difference. The fire-tube boiler is reliable and can be handled by those possessing comparatively little knowledge of engineering. Its chief defect seems to be the disastrous results following an explosion. The water-tube boiler, on the other hand, is safe, and suited to higher pressures, but requires greater care in management.

Before discussing these boilers in detail, let us consider briefly the salient points.

Safety. Probably the greatest advantage claimed for the water-tube boiler is its safety. The boiler contains much less water than does the flue or tubular boiler and the water is divided into small masses, thus minimizing serious results in case of rupture. On account of the shape and arrangement of parts, the circulation is usually good, and no part exposed to the fire can be uncovered while there is any water in the boiler. The tubes cannot become overheated until the boiler is empty and with an empty boiler there cannot be a serious explosion.

Rapidity in Raising Steam. The many small streams into which the water is divided as it passes through the furnace greatly facilitate the absorption of heat. Because of the small streams and the rapid circulation, the water is converted into steam in a very short time. Several hours (usually five to seven) are required to raise steam to working pressure in a tubular boiler, while in many water-tube boilers, steam can be raised to over 200 pounds pressure in less than half an hour.

Durability. Most water-tube boilers are so designed that no seams are exposed to the fire or hot gases. The seams are the weakest part of a boiler, and as strains due to unequal expansion concentrate at such points, leaks or even ruptures are liable to occur. In the water-tube boiler, the joints between tubes and tube sheets are not in the direct path of the hot gases.

Loss of Heat. The loss of heat will evidently be reduced to a minimum if the heating surfaces are such that the heat readily passes through the water. The small diameter of the water tubes (2 to 4 inches) allows the use of thin metal which does not hinder the transmission of heat. The rapid circulation in the water-tube boiler prevents the accumulation of sediment which is a poor conductor of heat. Still further, dust and dirt do not readily collect on the convex surface of water tubes, but the *inside* of fire tubes soon becomes choked with soot unless cleaned frequently. See Fig. 40.

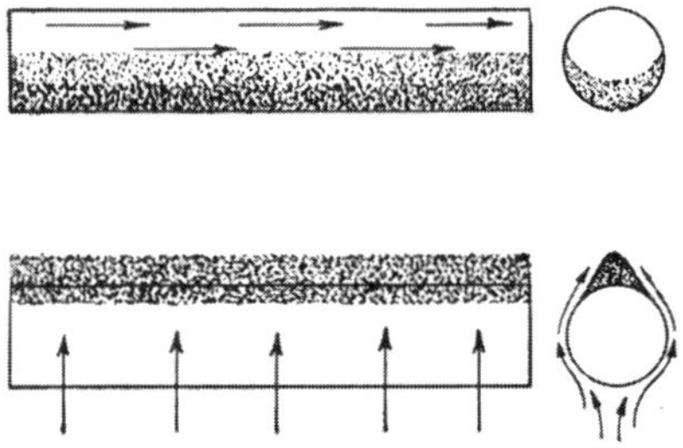

Fig. 40

Less Weight. It is a well-known fact that a cylinder of large diameter must be much thicker than one of small diameter when the internal pressure is the same. The thickness of the shell of a fire-tube stationary boiler is not excessive, because of the moderate diameter; but in the return-tube marine boiler, the shell plates for 250 pounds pressure would be about $1\frac{7}{8}$ inches thick. The difficulty of working such thick plates and their great weight render the cylindrical boiler unsuitable for high pressures. The small tubes and drums of the water-tube boiler may be made quite thin even for very high pressures. In general, it may be said that for the same capacity and pressure, the weight of a water-tube boiler is only about two-thirds that of a fire-tube.

Classifications

Many attempts have been made to classify water-tube boilers. By some writers a classification based on circulation, or on the principle of operation, is claimed to be superior to any division according to construction. Therefore, they divide them into classes as follows—boilers with limited circulation; boilers with free circulation; boilers with accelerated circulation.

In the first part of this Instruction Paper is given a classification according to features of construction. No classification is altogether satisfactory because boilers overlap into other divisions; a water-tube boiler may be sectional, of the double-tube type, have horizontal tubes, straight tubes, and free circulation. In order to have some sort of classification, and as no discussion will be entered into regarding relative merits, the classification given on pages vii and viii will be here adopted and followed as closely as conditions will permit.

Water-tube boilers are divided into two great classes—horizontal and vertical. Under these heads come sectional and non-sectional, straight-tube and curved-tube, and single-tube and double-tube. If the tubes are nearly horizontal, such as is the case of the Babcock and Wilcox, Root, etc., the boiler will be called horizontal. If the tubes are vertical, or nearly so, as in the Wickes, Stirling, etc., the boiler will be classed as vertical.

Although most boilers can be classified as outlined on pages vii and viii, there are a few of such peculiar construction and arrangement that they must be placed by themselves under "Peculiar Forms." These are described without any further attempt at classification.

As it is impossible to discuss all makes of boilers, a few representative forms will be considered as types of their respective classes. No attempt will be made to choose any make as being the best, because many conditions must be considered in selecting a boiler. The boilers described, except in a few cases, are now used extensively in either stationary or marine work.

Horizontal Water-Tube Boilers

Babcock and Wilcox

Water Tubes Nearly Horizontal—Steam and Water Drum Horizontal—Straight-Tube—Single-Tube—Sectional

Construction. This boiler consists of a large number of lap-welded, wrought-iron, 4-inch tubes connected to each other and to a horizontal steam and water drum. The arrangement of the parts is shown in Fig. 41 which is a side view of a much-used form of this boiler. Each tube is expanded into a forging of the form shown in Fig. 42.

The tubes in a vertical row enter one piece and this vertical row is independent of the others, as shown in Fig. 43. Thus it is readily seen that this is a sectional boiler. Fig. 43 shows also the "staggered" arrangement of the tubes. In the back side of the front header, and in the front side of the rear header, holes are drilled into which are expanded the water tubes. In the front side of the header a flanged hole opposite each tube is fitted with a hand-hole plate. The details of construction are shown in Fig. 44. The tops of the headers are connected to the steam and water drum by short tubes and the same construction is used for connecting the mud drum to the rear header.

Operation. The grate is at the front end of the boiler under the higher end of the tubes. The hot gases from the fire are guided by division plates and bridges, so that after rising from the grate they pass between the tubes to the

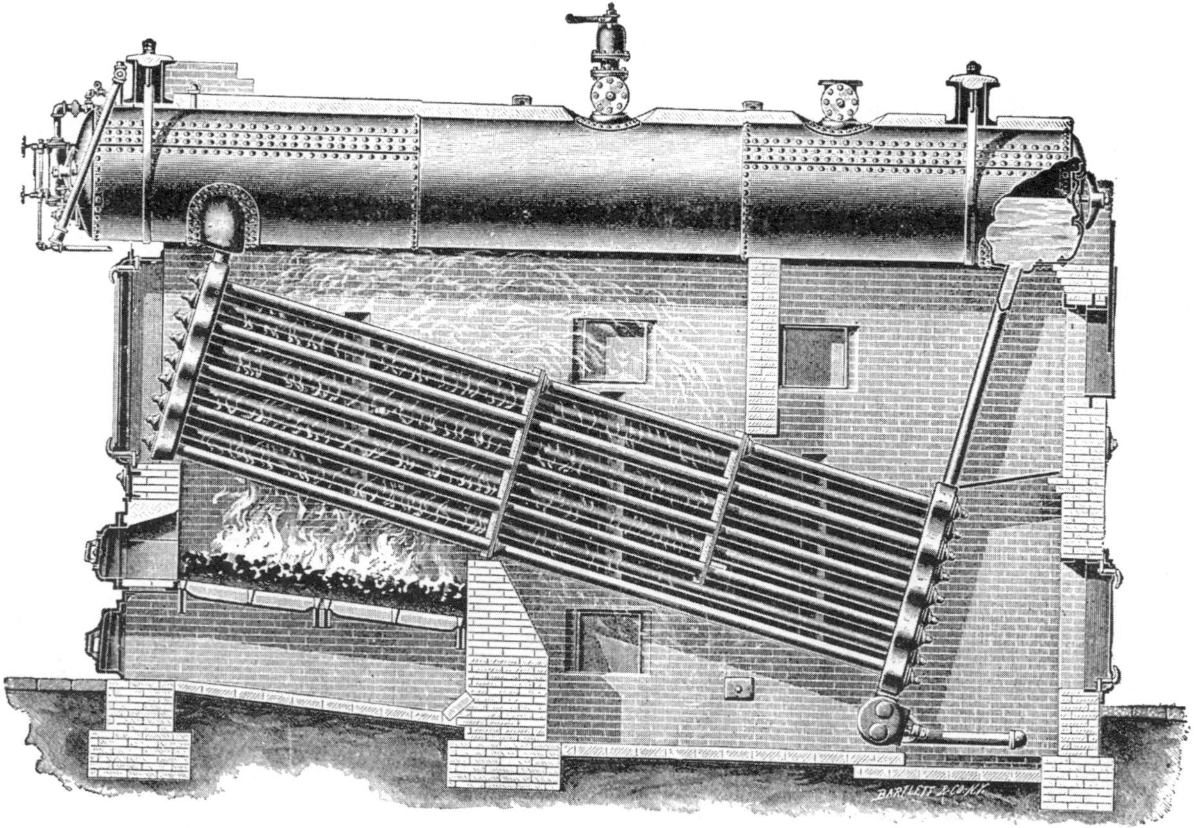

Fig. 41

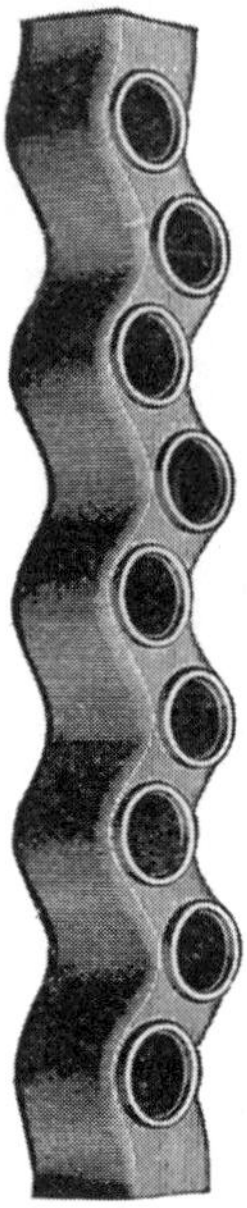

Fig. 42

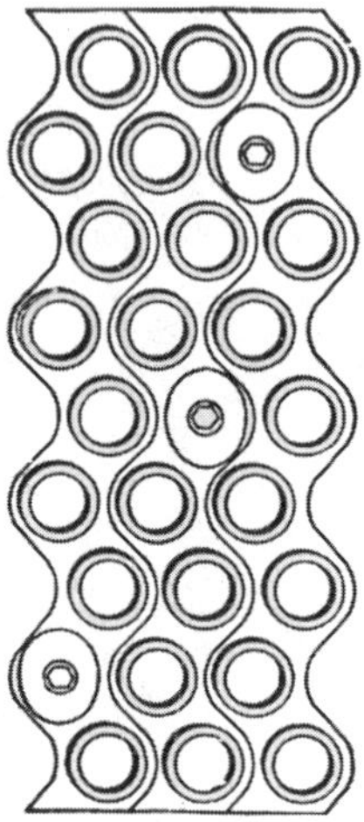

Fig. 43

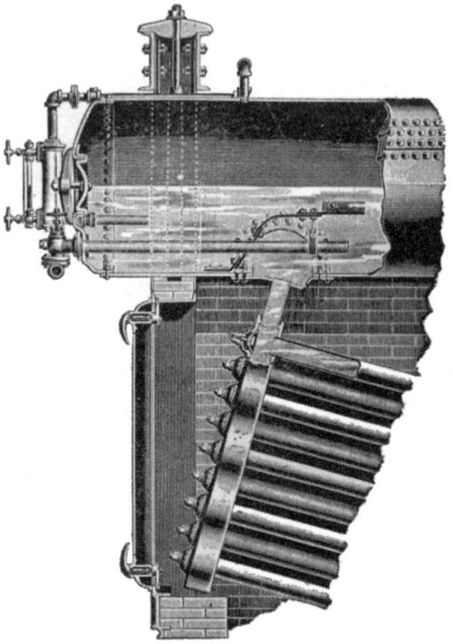

Fig. 44

combustion chamber, which is under the steam and water drum; the gases then pass downward among the tubes, and after rising a second time pass off to the chimney. In this way, the direction of the currents of hot gases is at all times almost at right angles to the tubes, thus impinging upon them instead of passing parallel to the heating surfaces, as in the case of fire tubes. As the gases impinge three times against the staggered tubes, the heating surface is very efficient.

Circulation. The feed water enters the steam and water drum through the pipe shown in Fig. 44. It is thus heated before it mixes with the hot water in the boiler. As the water in the tubes becomes heated, it rises to the higher end where it is partly converted into steam; a column of water and steam rises through the header to the drum in which the steam and water become separated. The cooler water at the rear of the steam and water drum flows down into the lower end of the tubes and as it becomes heated rises. Thus there is a continuous circulation.

Steam is taken from the rear end of the steam and water drum. The solid matter in the water is not deposited on the tubes because of the rapid circulation; it falls to the mud drum from which it is blown out.

The marine form of this boiler has a cross drum, that is, the drum is at right angles to the tubes instead of parallel to them. It is similar in form to the cross-drum types used for stationary work. This form is used in case there is not sufficient head room.

Root

Water Tubes Nearly Horizontal—Steam and Water Drums Horizontal—Straight-Tube—Single-Tube—Sectional

The above brief outline indicates that the Root water-tube boiler is, in its main features, like the Babcock and Wilcox. In fact the difference is in detail of construction only. Fig. 46 shows the general appearance when a part of the brickwork is removed. It will be seen that there is a large steam drum (cross type) at the top in addition to the small steam and water drum over each section.

Construction. The Root water-tube boiler is composed of 4-inch lap-welded wrought-iron tubes. These tubes are expanded into cast iron headers as shown in A, Fig. 45. A vertical section is formed by placing one pair upon another as shown at B, Fig. 45.

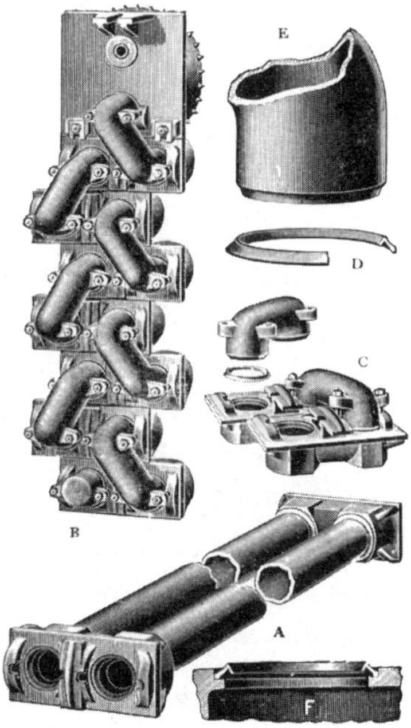

Fig. 45

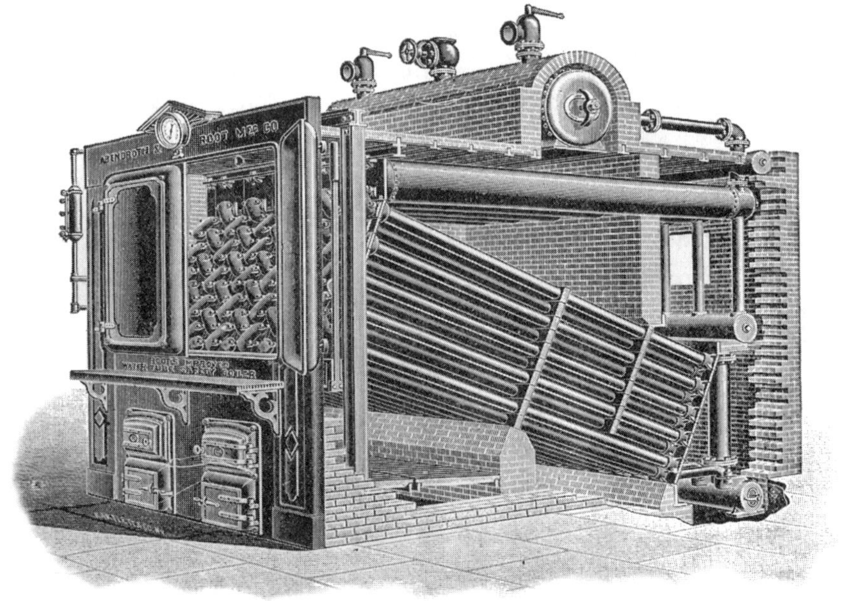

Fig. 46

One tube of each pair is connected to one above it by a flexible bend, by means of which is obtained an uninterrupted circulation from the bottom to the top of the section. A metallic packing ring (see C, D, and E, Fig. 45) insures a tight joint between the bend and the header. F, Fig. 45, shows an enlarged end of a bend.

To form the boiler several of these vertical sections are placed side by side. These vertical rows are not rigidly connected because the lower tubes being nearer the fire expand more than those above.

Circulation. Each section has its overhead drum into which the water and steam are discharged from the tubes. At the rear of the boiler and at the end of each steam and water drum, a vertical pipe leads to a cross drum beneath; this drum is a common reservoir for all the sections. The feed water enters this drum and meets the hot water coming from above. The mixing of the water results in a temperature which prevents any trouble from unequal expansion.

The cross drum (reservoir) is also connected by vertical pipes to another drum which is below and parallel to it; this is the mud drum. From the feed reservoir, the mixture of feed and circulating water descends to the mud drum in which the solid impurities are left. The circulating water then flows from the top of the mud drum into the lower end of the tubes. As these tubes are surrounded by hot gases, the water becomes heated and rises through the tubes to the steam and water drums. This heated water contains bubbles of steam which leave the water and collect in the steam drum. The water flows through the steam and water drum and descends to meet the entering feed water. The water level is at about the middle of the steam and water drums.

The hot gases from the fire pass among the tubes three times in practically the same manner as in the Babcock and Wilcox boiler.

Worthington

Water Tubes Nearly Horizontal—Steam and Water Drum Horizontal—Straight-Tube—Single-Tube—Sectional

Construction. This form of boiler is much the same in principle and operation as the Babcock and Wilcox boiler, but the parts are differently proportioned and arranged; see Fig. 47. The furnace extends under the entire boiler, and the tubes are set over it close together in oppositely inclined series. No flame walls or baffle plates are used.

Boilers up to 125 H.P. are usually made to fire at the end as shown in Fig. 47, in which the tubes extend across the furnace viewed from the front, and the steam and water drum is at right angles to the tubes. In the *side-fired* boiler the tubes extend from front to back, and the steam and water drum from side to side; this arrangement is better adapted for large units and for setting in battery. The tubes of each vertical row are expanded into straight headers which contain seven or eight tubes. See Fig. 48. Opposite each tube is a hand

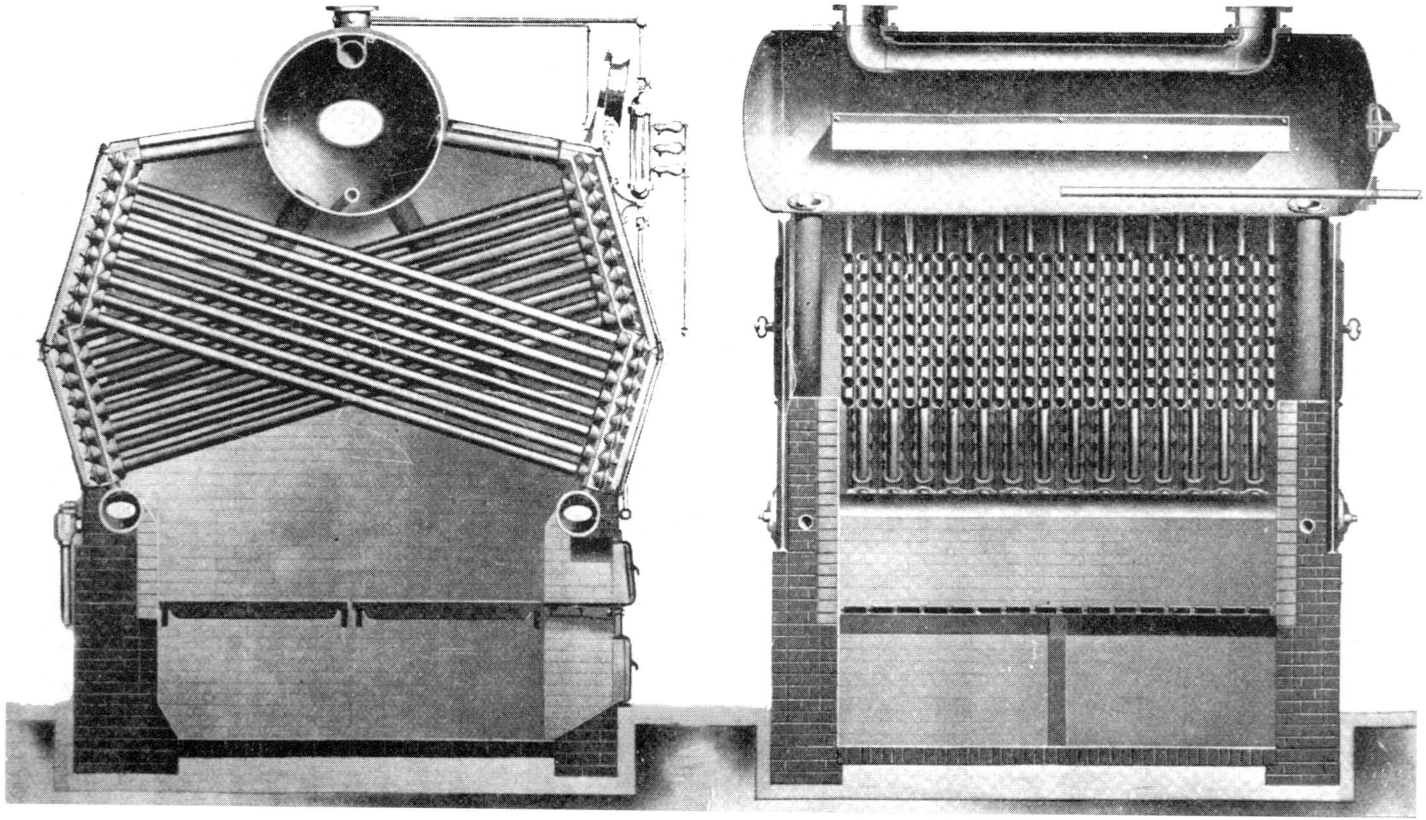

Fig. 47

hole. These headers are arranged close together, forming the boiler enclosure.

Circulation. The feed water enters the steam and water drum and the circulation carries it down to the mud drums through large circulating tubes which are outside of the furnace. See Fig. 48. From the mud drum it enters the lower series of headers and rises through the inclined tubes over the fire into the upper headers. The water, now containing bubbles of steam, enters the steam and water drum by means of short tubes shown in Figs. 47 and 48.

The covering for this boiler is an iron casing, no brick being used except to enclose or line the furnace.

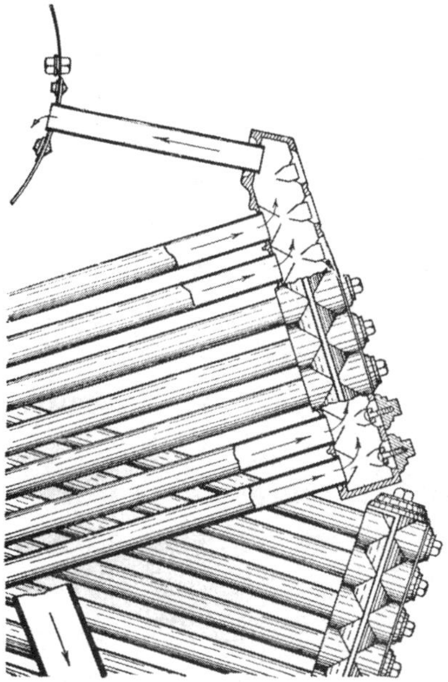

Fig. 48

Fig. 49

Heine

Water Tubes Nearly Horizontal—Steam and Water Drum Parallel to Tubes—Straight-Tube—Single-Tube—Non-Sectional

Construction. The Heine water-tube boiler is not a sectional boiler. Instead of being expanded into small headers grouped to form a boiler, all the tubes are expanded into the inside plates of a water leg at each end. The construction of this water leg is shown in Fig. 50. It is composed of two parallel plates flanged and riveted to a butt strap. The plates are strengthened by short hollow screw stays similar to those used in the water-leg construction of fire-box boilers. At the top, the water leg is curved and joined to the steam and water drum by riveting. Opposite each tube is a hand hole for cleaning or replacing a defective tube.

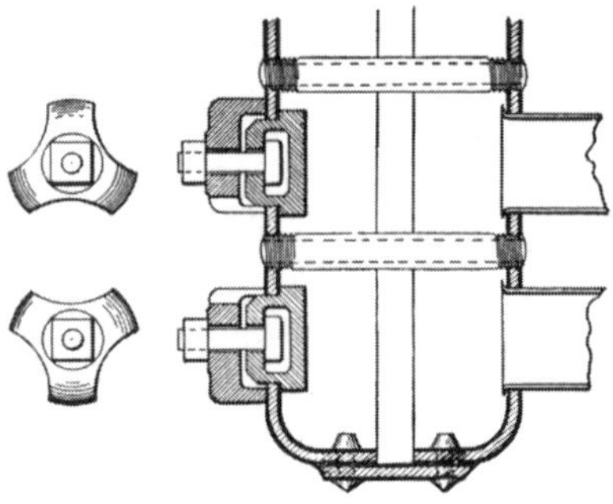

Fig. 50

Circulation. The feed water enters at the front of the steam and water drum and flows into the mud drum D, from which it passes to the rear header with much less velocity. The water is warmed while passing through the pipe leading to the mud drum, and as it flows slowly through the mud drum it deposits its sediment. The accumulated sediment is blown off by means of the blow-off pipe N. The water, as it becomes heated in the mud drum, rises and passes to the front of the mud drum, from which it flows in a thin sheet to the rear of the steam and water drum and to the rear water leg. From the rear water leg, it enters the tubes in which it is partially converted into steam. The mixture of steam and water enters the higher end of the drum from the water leg, and as there is but a thin layer of water in the steam and water drum, the steam readily rises through it. A deflection plate prevents water from being carried to the perforated steam pipe A.

The flow of hot gases from the fire is directed by light tile placed on the upper and lower rows of the tubes as shown in Fig. 51. The hot gases flow nearly parallel with the tubes instead of across them as in the Babcock and Wilcox.

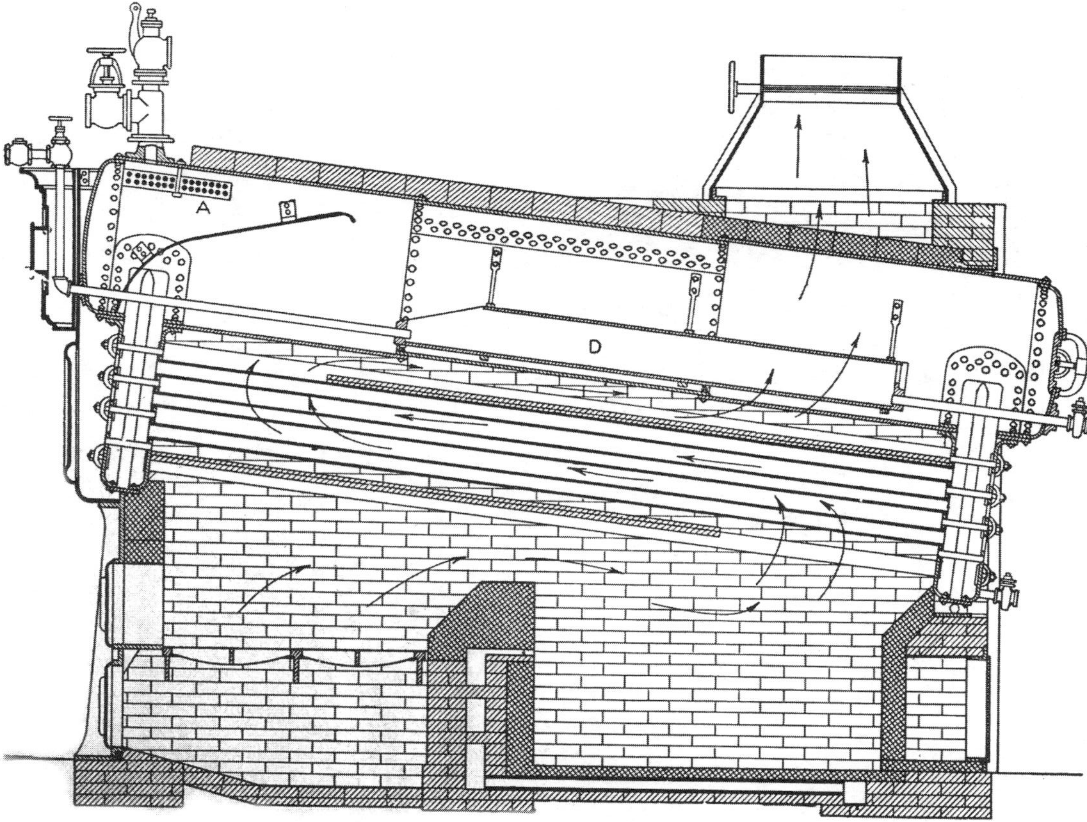

Fig. 51

Atlas

Water Tubes Nearly Horizontal—Steam and Water Drums (Cross Type) Horizontal—Straight-Tube—Single-Tube—Non-Sectional

This make of water-tube boiler does not need a full description as Figs. 52 and 53 show both the general arrangement of parts and the details of water-leg construction. The general description of the Heine boiler is applicable to this type. A few points of difference, however, should be pointed out.

Fig. 52

There are three drums running crosswise the tubes. The front and rear drums are made of the same plates as the water legs. This is shown in the illustrations. The reasons for this method of construction are that it gives a "throat" area of about 80 percent of the area of the leg and prevents all seams from coming in contact wit the furnace gases. The two drums are connected on the water line by equalizing tubes. In the rear drum, a water purifier receives the feed water which passes down the rear leg, then through the tubes to the front leg. In the last portion of the tubes, much of the water is converted into steam which flows through the front drum and the superheating tubes to the small upper drum. The water flows through the equalizing tubes to the rear drum and joins the current of feed water.

The front drum is made 36 inches in diameter, the middle drum 24 inches, and the rear drum 42 inches. The tubes are 4 inches in diameter and the longest are 18 feet in length.

Fig. 53

Mosher

Water Tubes Nearly Horizontal—Steam and Water Drums (Cross Types) Horizontal—Curved-Tube—Single-Tube—Non-Sectional

The chief differences in appearance between this boiler and those already described are shorter tubes, making a more compact boiler, and the curved tubes. This type is more often used in marine than in stationary work. The boiler consists of a large steam and water drum connected to a smaller water drum by slightly curved tubes. The steam drum is supported by two large circulating pipes (one at each end) which are connected by other pipes to the water drum. Thus the circulation is down these pipes and along the pipe at the bottom (see Fig. 55), up to the water drum and from thence to the steam drum by the tubes which are in contact with the hot gases.

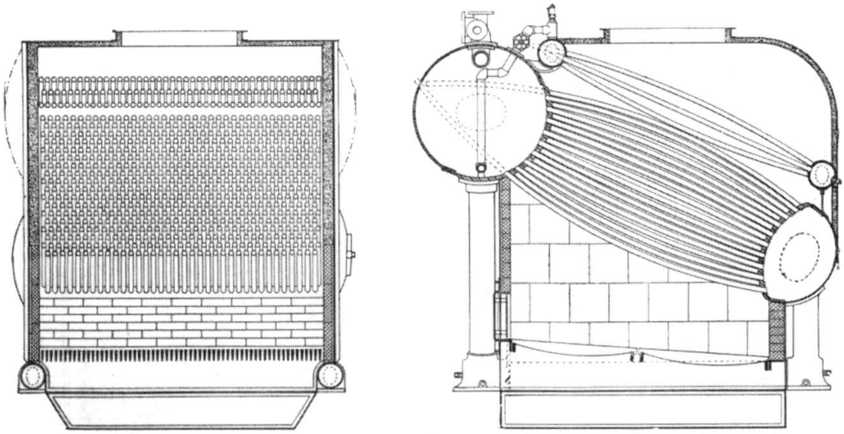

Fig. 54

The feed-water heater, shown in Fig. 54, consists of two small drums connected by tubes. The parallel dotted lines in the steam drum of Fig. 54 show how tubes are removed and replaced. Fig. 55 shows the row of plugs for this purpose.

Fig. 55

These plugs are illustrated in Fig. 56. Each plug is a conical-headed bolt, having a short piece of copper tube, a washer, and a nut. The conical head and the copper tube are inserted in the hole until the washer is in contact with the outer surface of the drum. The nut is then screwed up, thereby flaring the end of the copper tubing as shown. The steam pressure on the conical head increases the tightness of the joint.

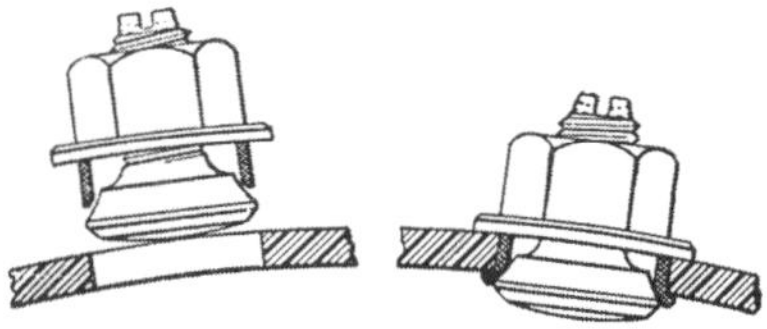

Fig. 56

Thornycroft-Marshall

Water Tubes Nearly Horizontal—Steam and Water Drum (Cross Type) Horizontal—Curved-Tube—Single-Tube—Non-Sectional

The Thornycroft-Marshall non-sectional boiler consists of a large horizontal steam and water drum, a vertical water box or header, and the generating tubes. Like the Mosher, the tubes are curved slightly, but the header is a distinct difference.

The general features of construction are shown in Fig. 57. The steam and water drum, sometimes called the separator barrel, is simply a cylinder with dished ends. The water level is about one-third the diameter of the cylinder. The

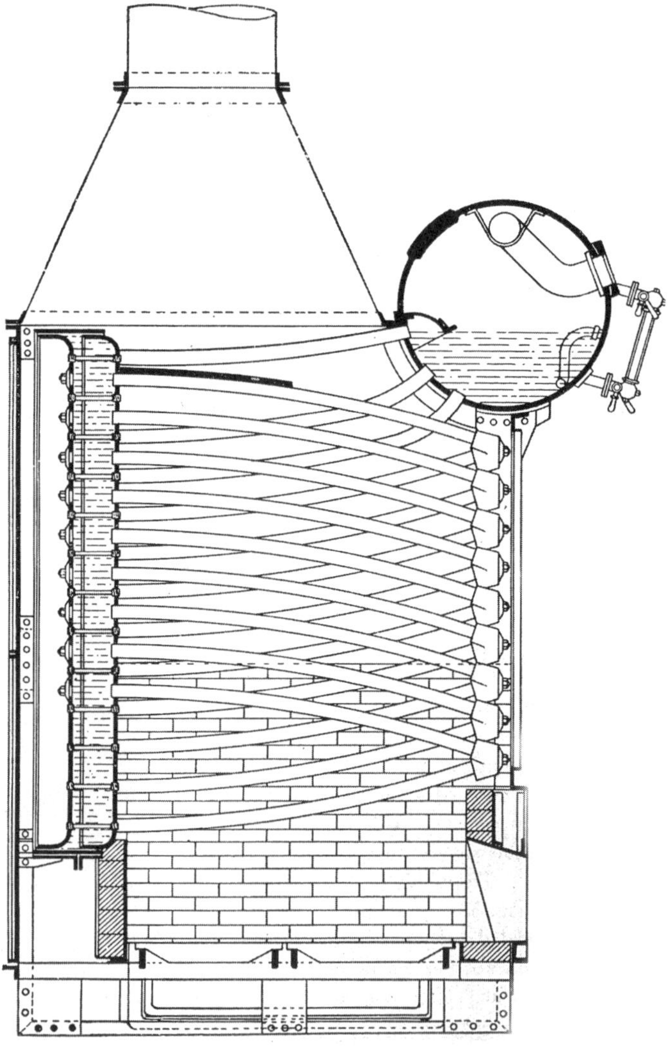

Fig. 57

tubes, which are $3^1/_4$ inches in diameter, are connected in pairs to a junction box at one end and to a water box or header at the other end. Thus each pair forms a unit, but the two tubes of the unit are not in the same vertical plane. The upper tube enters the header as high as possible and the lower ones enter low down, thus giving considerable upward slope. From near the top of the water box, three rows of tubes lead to the separator barrel as shown in Fig. 58. The water box is very simple, the flat plates are stayed by short hollow screw stay-bolts. The junction boxes are not restrained in any way; this construction, combined with the slight curve of the tubes, allows free expansion. The slight curve also allows the tubes to enter the separator barrel and the water box at right angles so that they may be expanded in place.

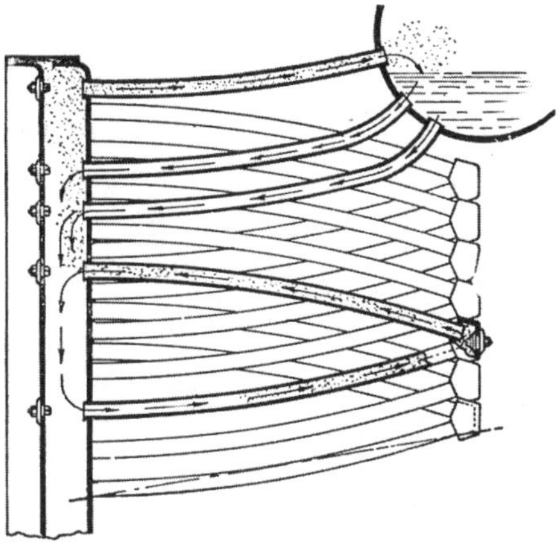

Fig. 58

Circulation. The feed water enters the steam and water drum and then passes to the water box through the two lower sets of tubes. See Fig. 58. The water enters the lower ends of the various pairs of tubes, as shown in Fig. 58, and rises in the tubes while in contact with the hot gases from the furnace. The mixture of steam and hot water then enters the header from which it passes to the steam and water drum by means of the highest row of tubes. The difference in height of the two tubes of a unit insures good circulation. A baffle plate prevents the water from splashing to the steam pipe.

The hot gases pass upward among the tubes which cross so frequently that they take almost all the heat from them.

Niclausse

Water Tubes Nearly Horizontal—Steam Drum Horizontal—Straight-Tube—Double-Tube—Sectional

The boiler differs essentially from those already described in that it is of the *double-tube* type. In general, it consists of a number of elements which form a vertical header, to which tubes are connected. The tubes are set at an angle of about 6 degrees to the horizontal. Above the elements is a transverse steam and water drum which is in communication with the headers. The general arrangement of parts is shown in Fig. 59.

Construction. The interesting features of this type of boiler are the design and construction of the tubes and headers. To increase the circulation the principle of the "Field" tube is employed. In this construction, the outer, or generating, tubes ($3^1/_4$ inches in diameter) are closed at one end. Each generating tube contains an inner circulating tube which is $1^{19}/_{32}$ inches in diameter. This tube is open at both ends. The closed ends of the generating tubes are supported by resting in holes in a plate or rack at the rear of the boiler. The forward end of the circulating tube is attached to a cap which screws into the outer end of the generating tube. A recess in this cap provides a bearing for an arch bar which spans two tubes, keeps them in place, and is itself secured by a nut on a bolt which is screwed into the header. See Fig. 60.

The front end of the generating tubes is of peculiar shape. To allow the water to enter the circulating tubes, and to fasten the tubes to the header without expanding them, each generating tube is provided at the open end with two cone-shaped portions; these are about eight inches apart. The first cone fits into a taper hole flanged outward in the front face of the header, and the second cone fits a similar hole in the rear face of the header. Both the holes and tubes are ground to the same size and taper. About midway between the cones, a third expanded portion occupies the tube

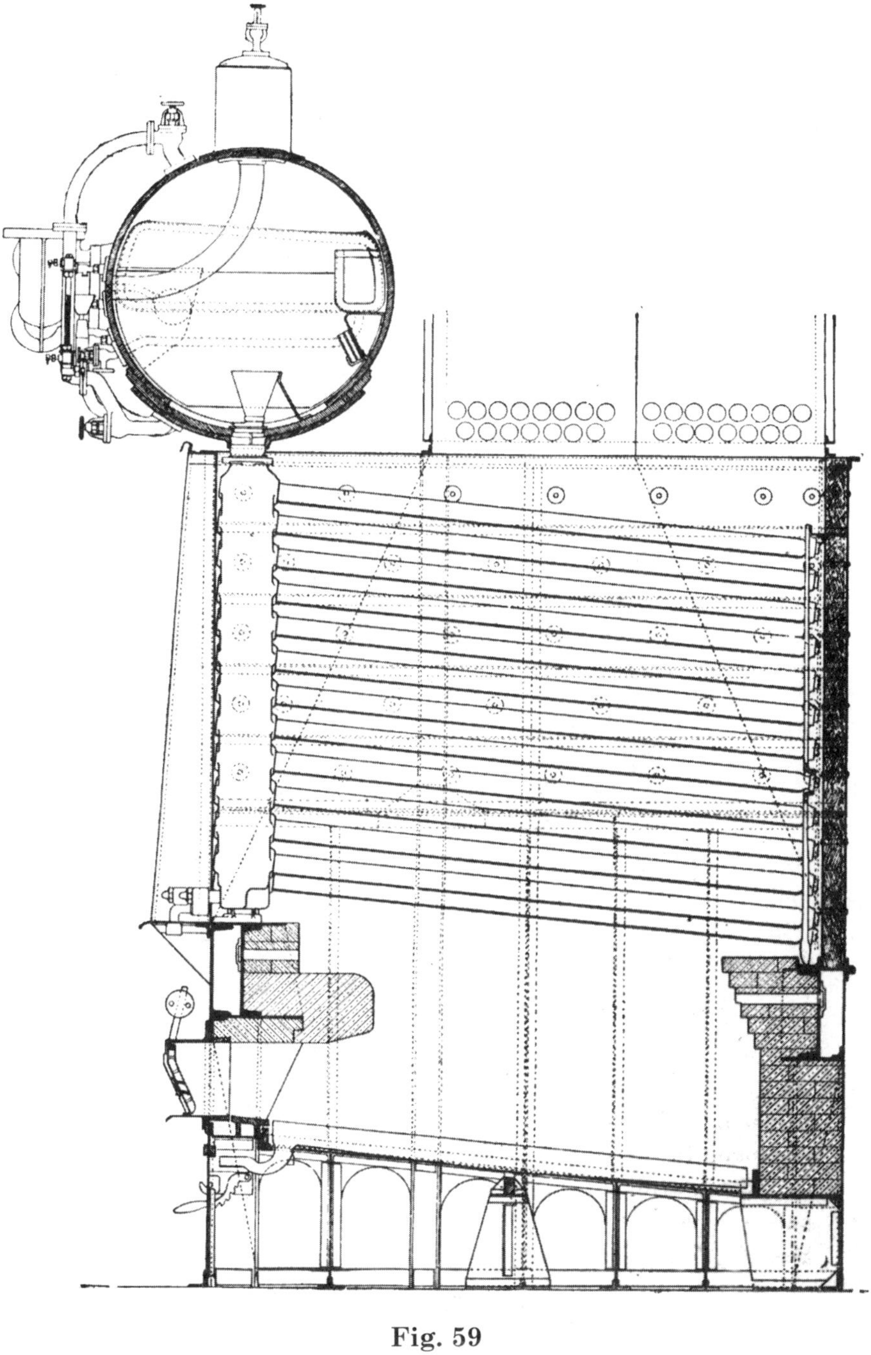

Fig. 59

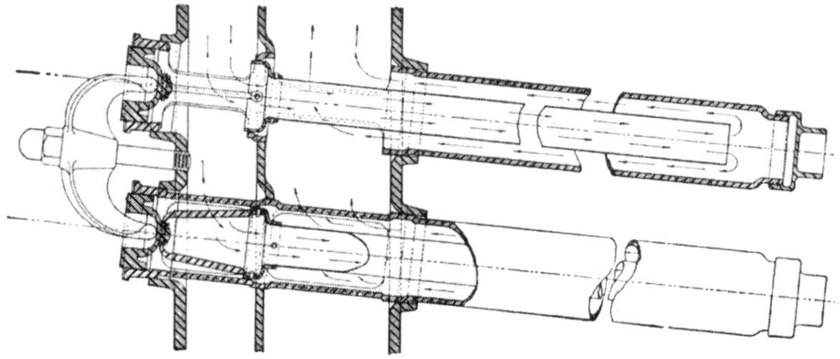

Fig. 60

hole in the diaphragm or middle plate of the header. See Fig. 60. The portion of the tube within the header is called the "lantern." At this point the tube is cut away so that the water may freely enter the tube, the openings being above and below. In Fig. 60, the upper tube is in its normal position, but the lower tube has been turned through 90 degrees to show the construction.

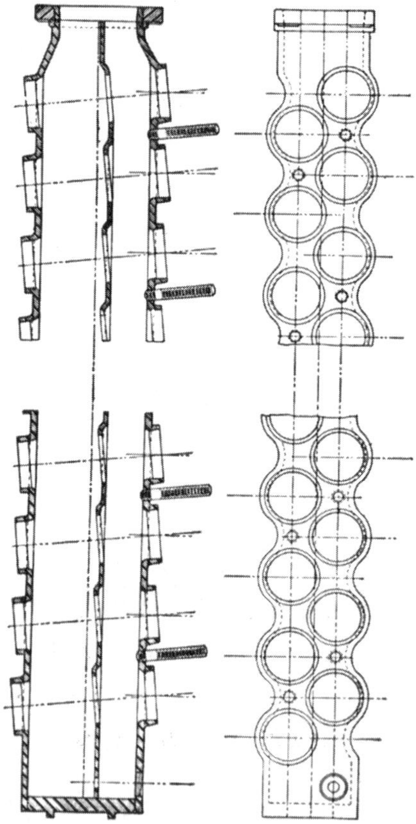

Fig. 61

To stand high steam pressures, the elements of the header are made of wrought steel and are sinuous in shape. Fig. 61 shows the shape of the header and the positions of the tubes. Each element contains 24 tubes in two vertical rows of 12 each. In the middle of the headers, there is a diaphragm for dividing the interior. The front passage serves as a "downcomer" for the water, and the rear is the "upcomer," or riser, for the mixture of steam and water.

The lower ends of the headers are closed, and the upper ends flanged to connect with the steam and water drum, which is 42 inches in diameter.

Circulation. Fig. 60 gives an idea of the direction of circulation. Water from the drum descends in the front compartment of the header, flows into the circulating tubes, *which communicate with the front compartment only*, and after flowing the length of the circulating tubes, enters the generating tubes. The water then comes back through the annular spaces in the generating tubes to the rear compartment of the header, because the *generating tubes communicate with the rear compartment only;* while in the annular space it is partially evaporated. The mixture of steam and water then rises to the drum.

Vertical Water-Tube Boilers

Wickes

Water Tubes Vertical—Straight-Tube—Single-Tube—Non-Sectional

Let us now consider a water-tube boiler having vertical tubes. Fig. 62 shows the general arrangement of the parts of the Wickes vertical water-tube boiler. At the top is a cylindrical steam and water drum into which the upper ends of the vertical tubes are expanded. At the bottom is a cylindrical mud drum of the same diameter as the upper drum. The tubes are straight and plum when in position; they are arranged in parallel rows with a clear space between rows to admit a small hoe to remove any soot that may accumulate on the tube sheet of the mud drum.

The tubes are divided into two compartments by heavy firebrick tile. The tubes in the section next the furnace are

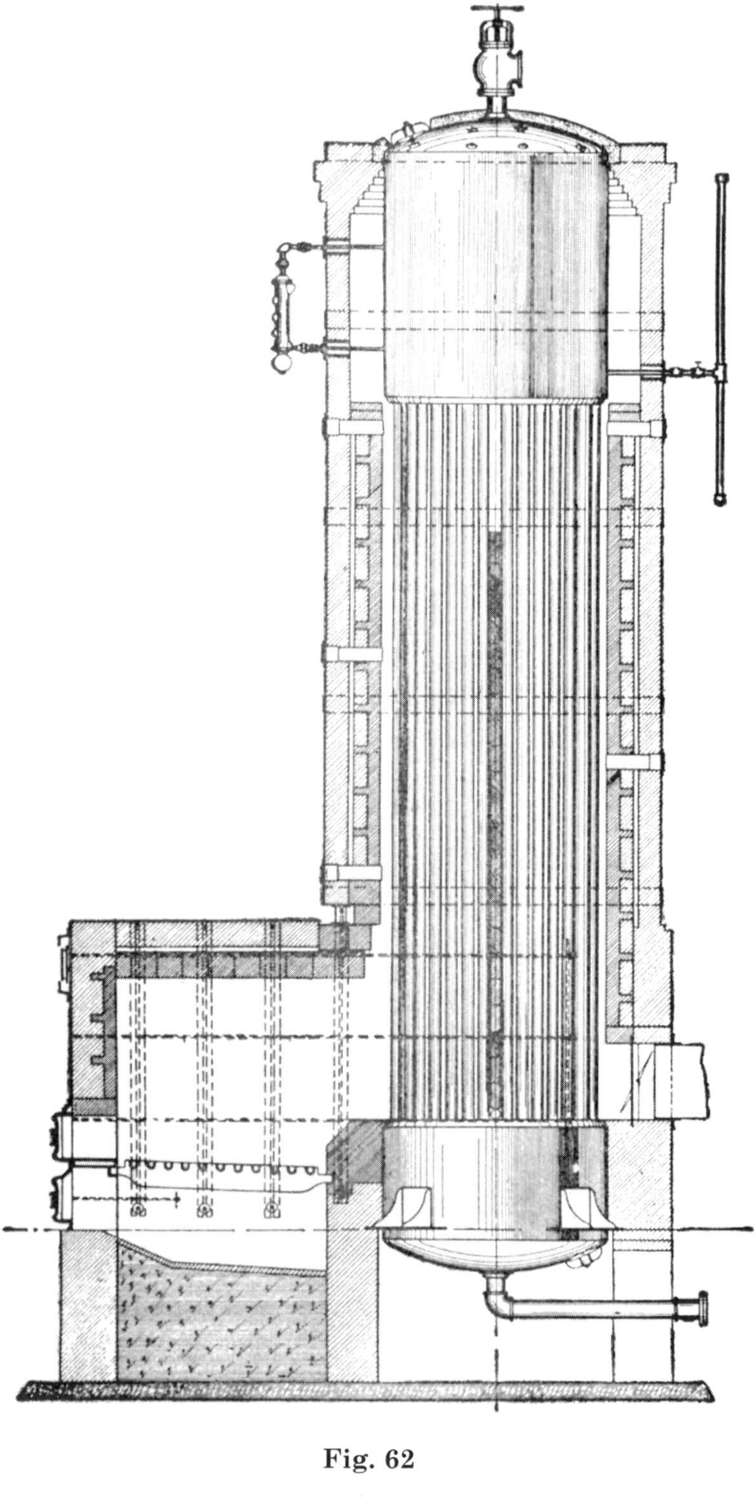

Fig. 62

called "risers"; those in the rear are the "downcomers," because the heated water rises to the steam drum through the front tubes, and the cooler water flows down those in the rear. The feed water is introduced into the upper drum. The direction of flow of hot gases is the same as that of the water. A baffle plate in the steam and water drum directs the water to the downcomers.

The furnace is external and built entirely of brick. The hot gases from the fire come in contact with the tubes without passing through a combustion chamber.

Cahall

Annular Steam and Water Drum—Water Tubes Vertical—Straight-Tube—Single-Tube—Non-Sectional

The Cahall vertical water-tube boiler consists of an annular steam and water drum, a cylindrical mud drum, and 4-inch vertical tubes. The generating tubes connect these two drums and are placed within the brick setting. An external circulating pipe also connects the two drums. As this pipe is filled with comparatively cool water and the generating tubes with a mixture of hot water and steam, the circulation is positive and rapid. The feed water enters the steam and water drum, flows down the external pipe to the mud drum and then rises in the generating tubes to the steam and water drum.

The fire is in a brick furnace at one side of the boiler as shown in Fig. 63. The hot gases rise among the tubes. The annular form of the steam drum makes the central space conical; in this space several deflecting plates, or baffles, cause the hot gases to flow out among the tubes. After heating the water in the tubes the hot gases pass through the opening in the steam and water drum coming in contact with the metal containing the steam. This thoroughly dries the steam and in many cases slightly superheats it.

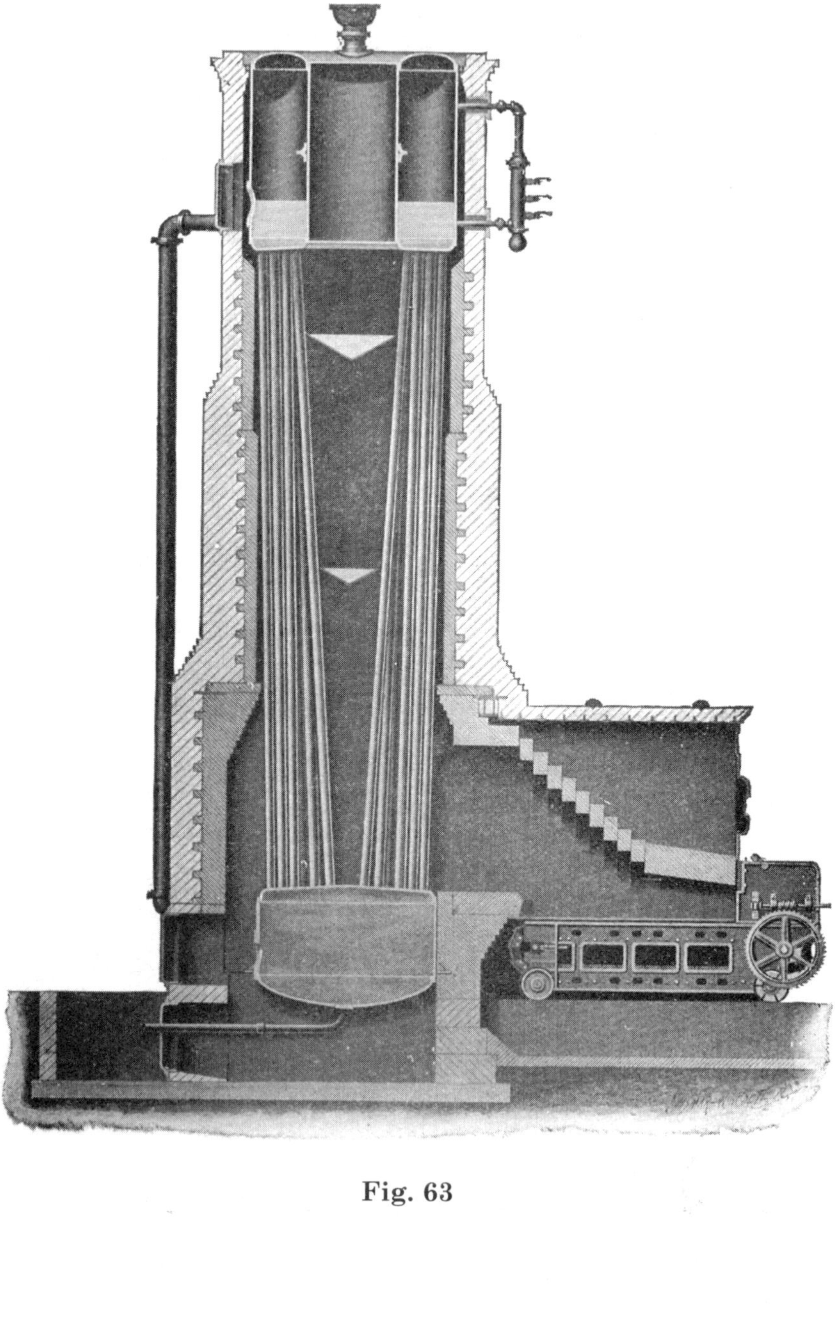

Fig. 63

The steam drum and also the mud drum are equipped with swinging manheads. The steam drum also has several handholes for use in removing and replacing tubes.

Stirling

Water Tubes Nearly Vertical—Steam and Water Drums Horizontal—Curved-Tubes—Single-Tube—Non-Sectional

The Stirling boiler, shown in Fig. 64, consists of three cylindrical steam and water drums at the top, and a mud drum at the bottom. The lower drum is connected to the upper drums by three sets of tubes which are curved slightly at the ends. The curved tubes allow for expansion and make it possible to have the tubes enter the drums radially.

The feed water enters the rear steam and water drum and coming in contact with the hot gases just before they enter the uptake, becomes gradually warmed. This heating causes most of the sediment to fall to the mud drum from

Fig. 64

which it may be blown out at intervals. The mud drum is protected from the intense heat of the furnace by the bridge wall.

Each set of tubes is separated from the others by partition walls or baffles of fire-brick tile so that the gases from the furnace pass along the entire length among the first set of tubes; they are then guided downward among the second set and after rising again among the tubes of the third set, escape to the chimney. By thus having a long passage a large proportion of the heat is taken from the gases before they go to the chimney. The fire-brick arch just above the furnace insures an even distribution of the gases and promotes combustion; the arch heats the entering air to a high temperature, thus reducing the liability of chilling the tubes by an inrush of cold air.

Steam is taken from the middle drum which is set a little higher than the others in order to obtain more steam space and drier steam. The boiler is surrounded on the rear and two sides by the brick setting; the front is of cast iron or of pressed steel. Numerous openings in the brickwork allow entrance for cleaning.

This type of boiler is flexible and adapted to cramped places as it can be made broad with little height or high with small floor area. All parts are either cylindrical or spherical in shape and of wrought metal. The curved tubes reduce the strains resulting from unequal expansion and contraction.

Milne

Water Tubes Vertical—Steam and Water Drum Horizontal—Curved-Tube—Single-Tube—Non-Sectional

This boiler (Fig. 65) is in many respects similar to the Stirling (Fig. 64), but an inspection of the two illustrations

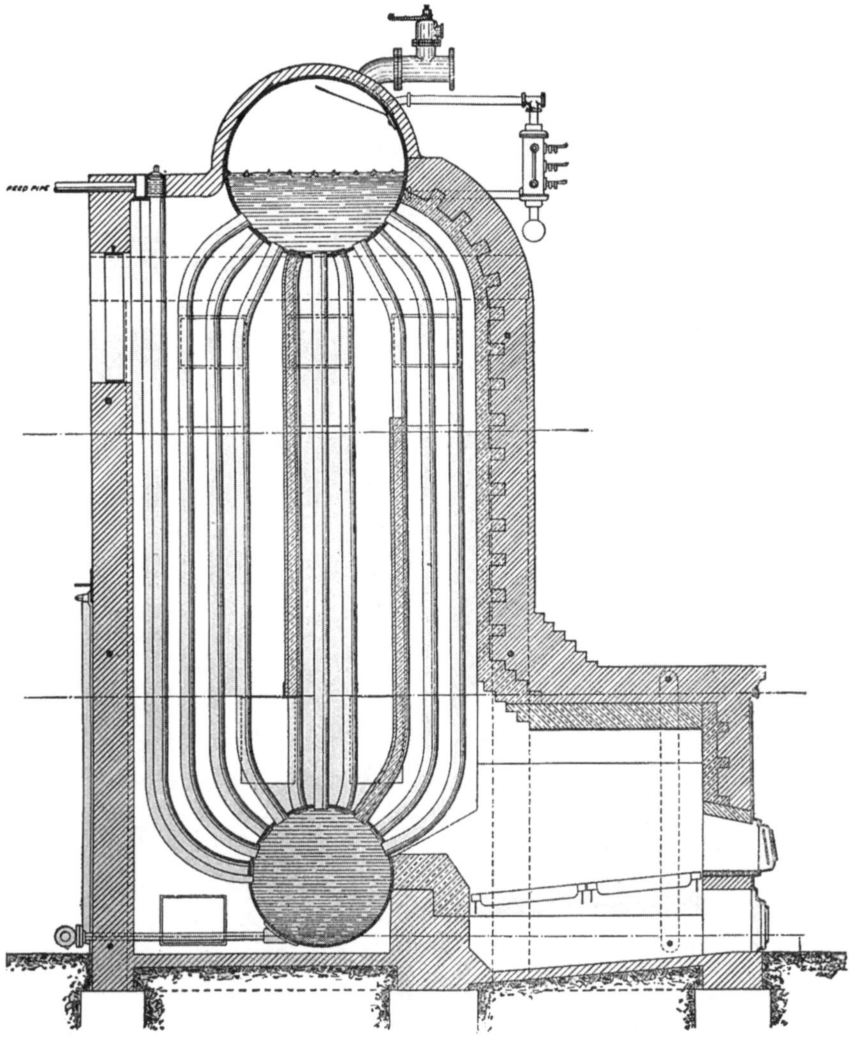

Fig. 65

will show several differences. In the Milne boiler there is but one steam and water drum and the tubes are vertical with a slight curve at the ends. The hot gases are guided by division plates or tile so that they traverse 65 feet of tube-heating surface before they enter the flue. The tubes, being vertical, do not become covered with fine ash, nor do they become clogged with sediment and scale.

Circulation. The feed water enters the row of tubes at the extreme left and flows downward to the mud drum. It then rises as it becomes heated in the hottest generating tubes and enters the steam drum as steam and water. This method of feeding keeps the cold feed water out of the steam drum, and as the cold tubes containing the feed are placed in the path of the escaping gases, but little heat escapes to the chimney.

Peculiar Forms

Hazelton or Porcupine

Water Tubes Horizontal—Steam and Water Drum Vertical—Straight-Tube—Single-Tube

The Hazelton water-tube boiler differs in many ways from the boilers thus far described. Like most water-tube boilers is consists of a steam and water drum and water tubes, but the central standpipe is vertical and the short horizontal tubes radiate from the central drum. According to our classification it is not a vertical water-tube boiler because the tubes are horizontal; also, it is not a horizontal boiler as in general appearance it is vertical.

The grate is circular and formed around the central drum which rests on a circular cast-iron foundation. Above the grate, the central drum forms part of the heating surface and is the steam reservoir; below the grate it is the mud drum, which may be entered by means of a manhole just below the grate. As shown in Fig. 66, the standpipe above the fire is provided with radial tubes. The appearance of these tubes gives the name "porcupine."

The standpipe is about three feet in diameter for large boilers. The tubes are about four inches in diameter, and two and one-half feet long, the number varying with the capacity of the boiler. The outer ends of the tubes are closed and hemispherical, and the inner ends expanded into the

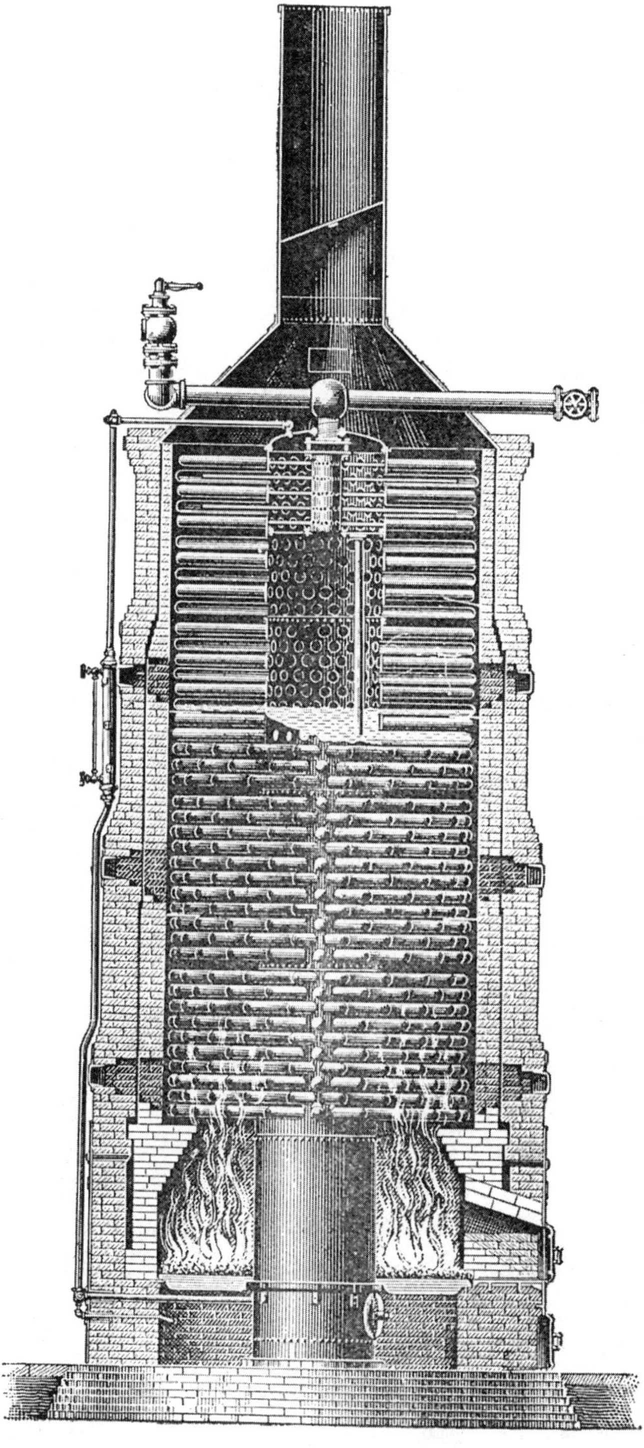

Fig. 66

standpipe. These tubes are free to expand and contract without bringing any strain on the boiler.

Steam is taken from the top of the central drum. To get dry steam, small pipes are inserted as shown in Fig. 67. The steam passes up into the small tube at the top of the standpipe and then through the small pipes to the ends of the generating tubes. It then flows back through the generating tubes to the annular space and from thence to the steam pipe. The feed pipe enters the mud drum and extends upward nearly to the water line; it then returns nearly to the level of the grate, terminating in a spraying nozzle.

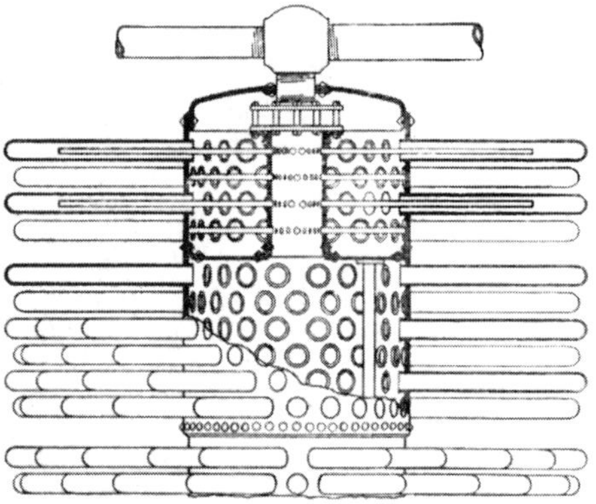

Fig. 67

This type of boiler may be enclosed in a brick setting as shown in Fig. 66 or by a sheet steel covering lined with fire brick.

Harrison

Sectional—Hollow Cast-Iron Spheres Instead of Tubes

All boilers thus far described have employed tubes as a means of dividing the water into small masses in order to make the heating surfaces more effective. In the Harrison Safety Boiler (Fig. 69) tubes are not used; instead, the water is contained in hollow cast-iron spheres, called units. These units, see Fig. 68, are arranged in vertical rows, called slabs, which are suspended side by side, about one inch apart, from an iron framework. The brickwork setting is merely a covering to keep the hot gases in contact with the units; it does not support the boiler, and can be repaired without disturbing the units.

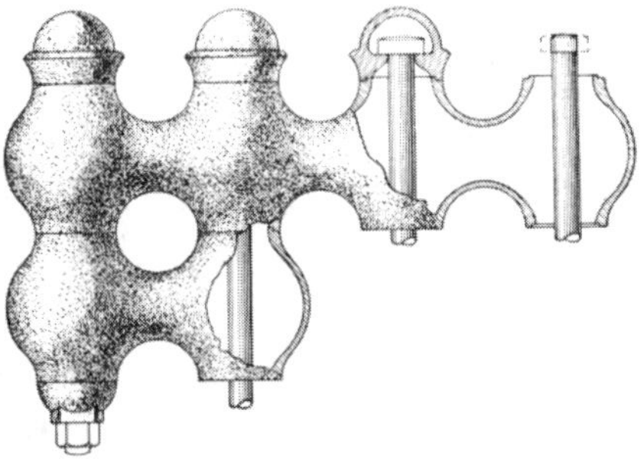

Fig. 68

The use of units in place of tubes combines great strength and a large heating surface. They are strong because small and spherical and on account of the division of the water into small masses, the heating surface is effective. The units are held together by long bolts which pass through the centers as shown in Fig. 68. The machined faces make a steam-tight joint without packing. This boiler requires the same fittings as other boilers.

The great advantage of this boiler is safety. From the construction, it is apparent that rupture cannot extend beyond the unit; thus disastrous explosions cannot occur. They are claimed to be durable, economical, rapid steamers, and easily handled. The capacity can be increased by merely adding more slabs.

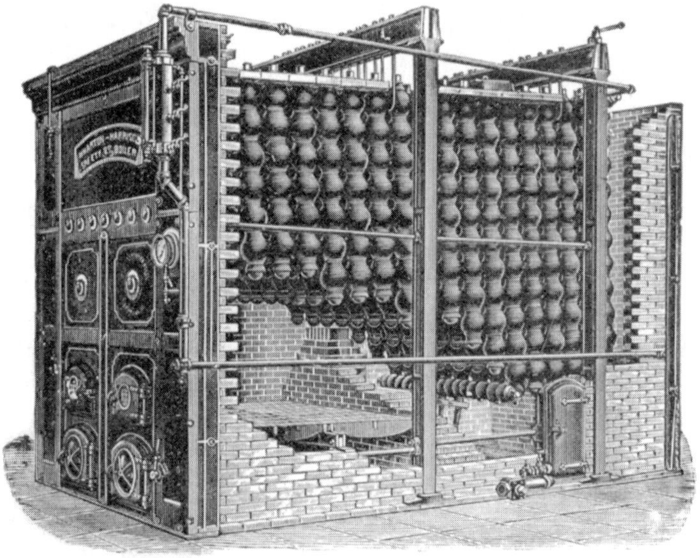

Fig. 69

Examination Paper

Read carefully: Place your name and full address at the head of the paper. Any cheap, light paper like the sample previously sent you may be used. Do not crowd your work, but arrange it neatly and legibly. *Do not copy the answers from the Instruction Paper; use your own words, so that we may be sure that you understand the subject.*

1. Why are water-tube boilers lighter than those of the fire-tube type?
2. In the fire-tube boiler, are the tubes large or small for forced draft? Why?
3. Make a sketch of a multitubular boiler and locate the important parts. Show by means of arrows the path of the hot gases.
   Note: The sketch should combine Figs. 12 and 13.
4. Name three boilers that have curved water tubes, are non-sectional, and have a steam drum of the cross type.
5. Describe briefly the Stirling boiler.
6. Classify, under the headings "sectional" and "non-sectional," all the water-tube boilers described in this Instruction Paper.
7. Trace the changes that occurred in the development of the horizontal multitubular boiler from the plain cylindrical.
8. What is the peculiarity of Beggs' Directurn Boiler?

9. What are the advantages of the double-ended return-tube boiler?
10. Trace the path of the hot gases in the Atlas boiler. For what purpose is the third drum?
11. In what way does the Niclausse boiler differ from all others here described? Trace the course of the water. Describe the construction.
12. Describe briefly the flue boilers—Cornish, Lancashire, and Galloway. Identify each by stating the peculiarities in a few words.
13. What is a Breeches-flued boiler?
14. Sketch the longitudinal section of a single-furnace single-ended return-tube boiler and name the principal parts.
15. Is the Babcock and Wilcox a sectional boiler? In what way does it differ from the Root?
16. What is the chief peculiarity of the Harrison boiler?
17. How does the Belpaire fire-box differ from the Wootten?
18. Why are vertical fire-tube boilers not economical? Describe the fire-box construction in the Manning boiler.
19. Describe the feed water heater used with the Mosher boiler. Is this a sectional boiler?
20. What two boilers belong to the class that may be briefly described as a straight vertical water-tube, single-tube, non-sectional?
21. Describe the Robb-Mumford boiler. Why is the furnace flue corrugated? Why is the lower shell inclined?
22. In what ways does the Hazelton boiler differ from the Wickes?
23. Why should the boiler shown in Fig. 34 be economical and at the same time occupy but little space?

24. Name all the boilers you can think of each one of which has all of the following characteristics: Water-Tube—Non-Sectional—Straight-Tube—Water-Tubes Horizontal or Nearly So—Steam Drum Horizontal.

25. Is the Worthington boiler sectional? What can you say regarding the extent of the furnace as compared with the tubes?

26. Discuss briefly some of the advantages claimed for water-tube boilers.

27. Why should one of the brackets supporting a multitubular boiler be placed on rollers?

28. In general, how are flat plates within a boiler prevented from bulging or collapsing? How are two flat or curved surfaces stayed when they are close together?

29. The Cahall boiler may be described as follows: Water Tubes Vertical—Straight-tube—Non-Sectional. Why cannot the Milne be placed in the same class?

After completing the work add and sign the following statement:

I hereby certify that the above work is entirely my own.

(signed)

SHIPCO® PUMPS

SHIPPENSBURG PUMP CO., INC.
P.O. BOX 279, SHIPPENSBURG, PA 17257
PHONE 717-532-7321 • FAX 717-532-7704
WWW.SHIPCOPUMPS.COM

SHIPCO® DEAERATOR SERIES

PRIDE • QUALITY • CRAFTSMANSHIP

Why a Deaerator? What Does It Do?

A deaerator is a special kind of boiler feed pump that minimizes the corrosiveness of the boiler feed water and pumps this water into the boiler as needed. It reduces the corrosiveness by heating the water, thus removing harmful dissolved gases (oxygen and carbon dioxide) by driving them out of solution.

Since corrosion, like other chemical reactions, increases with temperature, a higher rated deaerator is needed for higher temperature boilers. The temperature and pressure of boiling water rise simultaneously; it is therefore important to use the best type of deaerator for high pressure boilers to protect the boiler and the system from corrosion while minimizing boiler chemicals.

Lower priced, less effective deaerators can be a good compromise for low pressure boilers. Standard industry ratings are .005 for the highest rated, and .03 for the less expensive deaerators (Rating .005 = .005 cubic centimeters of oxygen remain per liter of feed water). This .005 rating is equivalent to about 7 parts per billion by weight, which is a minute amount of remaining oxygen. The .03 cc/liter rating represents about 44 parts per billion of oxygen in the water by weight.

Actual performance can vary a lot and depends on the force of the boiling water (temperature and turbulence), and the steadiness of the flow into the deaeration compartment (transients), as well as other factors which will be mentioned in "How a SHIPCO® Deaerator Works."

Good performance depends on hard boiling, therefore the generated steam must be highly condensed before the gases are vented to atmosphere to avoid loss of steam. Design and proper controls are equally important for the efficient operation of the deaerator. Properly operated deaerators require drastically fewer oxygen scavenging boiler treatment chemicals.

In summary deaeration achieves the following:

- Removes air (oxygen and carbon dioxide)
- Raises feed water temperature (reduces boiler shock)
- Improves heat transfer (air acts as an insulator, hindering heat transfer in the system)
- Saves money (limits chemicals, limits boiler retubing, saves return lines, heat exchangers & process equipment).

Fundamentals for Outstanding Deaerator Design

1) Temperature
2) Time
3) Turbulence
4) Thin Film
5) Transients
6) Venting
7) Vent condensing

Use 5 T's and 2 V's as an aid to remember these very important requirements.

Heat is required to raise the water **temperature** in your deaerator to full saturation temperature for the internal pressure of the deaerator. The temperature must remain at the boiling point to ensure oxygen removal.

Time is required to remove all traces of dissolved oxygen. The more time allowed in each step of the deaerating process the more effective the deaerator becomes.

Turbulence is required to vigorously scrub all the gas bubbles. Steam added at the bottom of the unit through the steam manifold assembly causes steam bubbles to rise and carry any remaining tiny air bubbles to the surface. This process helps overcome surface tension that retains dissolved gases.

A **thin film** is required to decrease the distance tiny bubbles must travel to be released. By reducing the time required to accomplish release, the quantity that can be handled by a given size unit is increased. This increase is accomplished by spraying the water through nozzles and then by continuing the process over a series of baffles.

By controlling all **transients** with modulating controls and allowing returns to enter a surge tank with pumps running continuously, temperature and capacity variations are minimized. Quick and sudden changes can cause air to redissolve very rapidly.

Venting of the liberated non-condensable gases with the escape of some steam to atmosphere must occur. If these gases were not allowed to escape, deaeration could not happen.

Lastly, **vent condensing** or adding a shell and tube vent condenser/preheater to the vent of a deaerator can heat low temperature water hotter before spraying, and condense nearly all of the steam from the vent, increasing the efficiency of the deaerator.

How a SHIPCO® Deaerator Works
Important Features

The principle on which a deaerator works is Henry's Law, expressed graphically on the next page. At the chosen pressure the amount of dissolved gases goes to zero when the water is at the boiling temperature for that particular pressure. Since it takes time for a bubble of oxygen to form and rise, boiling the water hard shortens this time. It is therefore possible to have a deaerator work at vacuum, atmospheric pressure, or a higher pressure. Atmospheric and low pressure (5 PSI) are most often used. SHIPCO® manufactures both of these deaerators. Advantages and disadvantages are as follow:

Atmospheric Advantages:

- No ASME Code tank expense
- No float drainer for overflow
- No safety valve
- Less steam loss at partial load
- No orifice in vent
- Shell & tube vent condenser/preheater advantage for larger temperature rise heating requirements

Pressurized Advantage:

- Faster response of pressure pilot

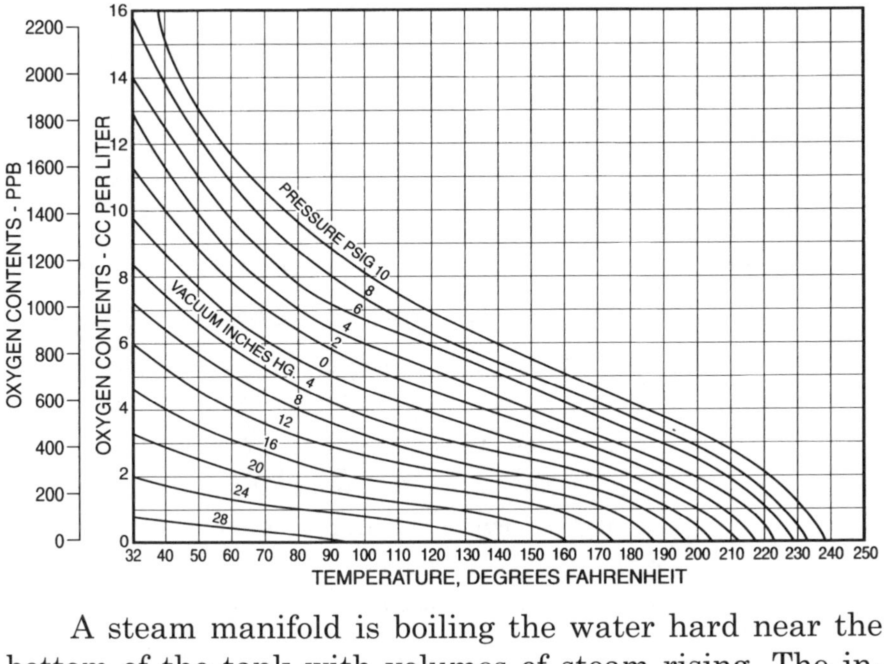

A steam manifold is boiling the water hard near the bottom of the tank with volumes of steam rising. The incoming water is sprayed into the top of the tank just below the vent, condensing the steam and heating the incoming water. It then cascades through the baffles further heating the water, condensing more steam, and releasing oxygen which travels up toward the vent with the steam. The water, then mostly deaerated, drops to the surface where the bubbling steam purges the last of the oxygen. The deaerated water is stored until the boiler calls for water and it is pumped from the bottom of the tank to the boiler.

Important Features:

1) A path in one direction for the steam to carry the oxygen toward the vent, effectively isolating it from the deaerated water.

2) Spring-loaded spray nozzles that provide a fine spray at varying flow rates for rapid heating and oxygen removal.

3) A series of baffles further providing for heating of water, condensing of steam, and release of oxygen, as well as providing the one-direction pathway.

4) A steam manifold with narrow slits to provide good distribution of fine steam bubbles to scrub the last of the oxygen from the stored water. It will operate relatively quietly above 160°F.

5) This scrubbing heater (a little above the bottom) assures the pumps are getting the most thoroughly deaerated water in the unit. Since heat rises, steam added at the bottom is more effective in heating the bottom as opposed to adding heat at a higher level. Furthermore, the rising steam bubbles are very effective in carrying any remaining tiny air bubbles to the surface.

6) Noticeably better deaeration can be obtained when flows increase or decrease gradually. Sudden increases and decreases, such as a large on and off flow from a condensate pump, will hinder performance unless a surge tank is used with modulated flows from the surge tank to the deaerator tank. A system with 80% to 100% make-up water

can operate efficiently without a surge tank as the water flow can be modulated from the domestic water supply, but larger percentages of pumped returns require a surge tank for best performance.

7) A shell and tube vent condenser/preheater, if added to the vent of the deaerator, increases the efficiency of the deaerator by heating low temperature water hotter before spraying, and condensing nearly all of the steam from the vent.

Deaerator Boiler Feed Pump Selection & Sizing

Selection is based on gallons per minute (GPM), pounds per square inch (PSIG), net positive suction head (NPSH) and receiver size.

Determine GPM

The evaporation rate of one boiler horsepower is .069 gallons per minute.

Other conversion equivalents: One boiler horsepower equals 33,475 BTU/hr. or 34.5 lbs./hr. or 139.4 sq. ft. EDR.

Boiler feed pumps for on-off operation are sized at two (2) times this evaporation rate.

Boiler feed pumps for continuous operation (generally 15 motor HP and larger) are sized at one and a half ($1^1/_2$) times this evaporation rate. In addition, extra flow for recirculation with deaerator boiler feed pumps may have to be added. The SHIPCO® centrifugal Model "P" and "D" pumps do not require any additional flow. SHIPCO® pumps with motors 5 HP and less have a bleed line that serves this function, and in pumps with motors $7^1/_2$ HP and larger a bypass orifice is used in a recirculation line for this purpose.

Boiler feed pumps are sized based on the maximum number of boilers each pump is feeding.

Determine PSIG

The deaerator boiler feed pumps are sized to overcome the **operating** pressure of the boiler + friction loss in pipe + valve loss + feed valve loss (if any) + stack economizer (if any) + vertical lift from pump to boiler + safety margin of approximately 10 PSIG. The amount of these values added together, are normally expressed in feet of head. To convert feet of head to PSIG, 2.31 ft. = 1 PSIG.

Generally, the feed valve loss is 20 PSIG and the stack economizer loss is 20 PSIG when estimating the pump discharge pressure. Stack economizer requires a continuously running pump in the system.

The standard rules of thumb are:

- High-pressure boilers running on–off from a boiler level controller add 20 PSIG to the operating pressure (not the design pressure).
- High-pressure boilers running continuously pumping through a modulating valve add 30 PSIG to the operating pressure (always better to get pressure drop through valve).
- High-pressure boilers running continuously pumping through a modulating valve and a stack economizer add 50 PSIG to the operating pressure.
- Low-pressure boilers (running between 1 to 15 PSIG) generally use pumps with a discharge pressure of 20 PSIG.

If the boilers run at more than one pressure setting (like a night setback), an additional pump(s) is needed to handle this pressure and the steam control regulator must be sized for this nighttime low-pressure setback.

Determine NPSH

NPSH stands for Net Positive Suction Head. The **available NPSH** is essentially the measure of how close the water in the suction passage of the pump is to boiling, with the attendant formation of steam within the impeller, thus diminishing the pump's performance.

Since we have a deaerator where the water is at the saturation point or boiling point, the **available NPSH** is at zero, located at the bottom of the steam manifold.

Various physical designs of pump have various **NPSH requirements**. In order for any pump to operate successfully, the NPSH **available** must be **greater** than the NPSH **requirements.** With a deaerator the only way you can increase the NPSH available is to elevate the tank a greater distance than the pump requires. For example, a pump with an NPSH requirement of 4 ft. must be elevated at least 4 ft.

Suction strainers hurt NPSH calculations since you can't measure the pressure drop through a strainer. In addition, if it works it will destroy the pump by causing it to run dry. For this reason suction strainers are **never** used with centrifugal pumps. Suction strainers are only used when turbine pumps are supplied since even a little dirt and debris will cause this style of pump to go bad due to the close tolerances within the design. The standard rule of thumb is to add one additional foot of stand elevation to compensate for this suction strainer.

Determine Receiver Size

The receiver size on a deaerator is based on the total load of all boilers being fed by the unit at any one time. The receiver size is generally based on 10 minutes of net storage when using a single compartment with returns.

If the system utilizes a surge tank with the deaerator, then the surge tank will be sized to handle the 10 minutes of net storage time required, with the deaerator being sized for only 5 minutes of net storage.

A deaerator without returns (100% make-up) requires only 5 minutes of net storage.

As demonstrated, the selection of the receiver size may vary based on the characteristics of the system.

Surge Tank Pump Selection & Sizing

When Is a Surge Tank Used?

The general rules of thumb are as follows:

On systems with **80% or more make-up** a surge tank is not **required**.

On systems with **more than 20% returns** a surge tank is **required** to achieve good deaeration.

What Is a Surge Tank?

A surge tank is another name for a boiler feed tank. It acts exactly like a boiler feed tank except that it feeds a deaerator in lieu of a boiler. With a surge tank the make-up water is added into this tank and blended with the return water to avoid shocking the deaerator with extreme temperature and capacity variations. In addition, the pumps on the surge tank **must run continuously**, pumping the water directly into the modulating transfer or the make-up valve on the deaerator. The second transfer pump on a SHIPCO® surge tank is a standby pump that is activated by a low-level switch on the deaerator (-2T or -2C Type Units). This standby pump runs automatically in case the lead pump fails or can't keep up.

A surge tank is <u>not</u> a condensate pump since a condensate pump turns on and off based on the water level in its receiver. When a condensate style unit is used as a surge tank, it defeats the entire purpose of a surge tank by allowing large variations in capacity and temperature into the system. The main purpose of the surge tank is to level out the transients or control the fluctuations in capacity and temperature so the deaerator runs as smoothly and effectively as possible.

If controlling these variations in temperature were not important, no need would exist to use expensive controls that modulate on the deaerator.

What Does a Surge Tank Look Like?

Since a surge tank is another name for a boiler feed tank, as mentioned earlier, the tank can take many shapes and forms. The tank can be made of stainless steel, close grained cast iron (with a 20-year warranty against corrosion failure) to prevent against corrosion failure, Plasite #7156 lined, or simply black steel.

The surge tank can be an integral part of the deaerator, like a two-compartment style (look under its specific tab), or be free standing by itself (called a two-tank or -2T system).

Where surge tanks are in a free-standing by itself situation (-2T systems), the free-standing surge tank can be elevated or mounted on the floor like many of the types throughout the entire catalog (see Reference Chart A below). Yes, the surge tank may be placed on the floor. This is possible because at 150, 180 or even 200°F the water temperature is low enough that NPSH is not a major concern. For example, with 194°F water you have 10.46 feet of NPSH available at sea level. If you look at the pump curves in the catalog under Model D or P, the pumps, if properly selected, are 2 ft., 4 ft. or 6 ft. NPSH at the best efficient point on the curve; hence, the NPSH available is greater than the pump NPSH requirement based on 194°F water in tank.

All floor-mounted units should be sold and specified with SHIPCO® bronze isolated valves that are factory tested for servicing the pump.

How to Determine Transfer Pump (Gallons per Minute) GPM

All deaerator units are rated in lbs/hr of steam. The transfer pumps on the surge tank units are sized based on this rating. Lbs/hr divided by 500 equals the evaporation rate in GPM for these pumps. The pumps are sized as follows:

If transfer pumps are feeding a deaerator on a free-standing by itself surge tank (-2T) system, the pump rate in GPM equals the evaporation rate or the total load rating on the deaerator. For example, if deaerator system is rated 10,000 lbs/hr, then each transfer pump should be rated for 20 GPM.

If transfer pumps feeding a deaerator with the surge tank are part of the complete unit like our two-compartment model (-2C), the transfer pumps are sized differently based on being an atmospheric or pressurized deaerator.

If a pressurized two-compartment deaerator unit is used (-2C), the pump rate equals the evaporation rate or the total load rating on the deaerator (as mentioned earlier).

If an atmospheric two-compartment deaerator is used (.005 DA-2C or .03 DA-2C), the pump rate in GPM equals the evaporation rate of the deaerator multiplied by 1.5. For example, if deaerator is rated 10,000 lbs/hr, then each transfer pump should be rated for 30 GPM (20 GPM x 1.5). This is to allow recirculation of water through the vent condensers. *In addition, this is the only type of surge tank where NPSH is a concern and pumps should have an NPSH requirement lower than the height of the stand to be safe.*

Recirculation for these continuously running transfer pumps may be required. The SHIPCO® Model P and D pumps have as standard a bleed line that does not require any additional recirculation when pumping liquids lower in temperature than the saturation or boiling point.

How to Determine the Pump Discharge Pressure PSIG

The surge tank pumps are sized as follows:

A) To overcome the operating pressure of the deaerator

B) Spray nozzles

C) Friction loss in pipe

D) Vertical lift between deaerator and surge tank

E) Safety margin generally 5 PSIG

F) Pressure drop associated with transfer valve

The amount of these values, or these values added together, is normally expressed in feet of head. To convert to pounds per square inch, or PSIG, 2.31 ft. = 1 PSIG.

Generally the surge tank is located beside the deaerator or when it is part of the deaerator itself; therefore, a transfer pump discharge of 25 PSIG is used as our standard since our standard transfer modulating valves are sized for 100% of the deaerator load with a 10 PSIG drop across valve.

How to Size and What Style of Make-Up to Use

Since the purpose of a surge tank is to gather all the returns from the system and mix the make-up water with the returns to blend the temperature, the make-up valve should be a standard close closing solenoid valve activated by a float switch in the tank. By using a modulating make-up valve you are wasting money on an expensive valve and controller that serve no purpose unless your system is larger than 75,000 lbs/hr.

The make-up valve capacity should be sized for 100% of the total load of the system so you have emergency backup. In addition, make sure that your city water supply pressure is adequate for the pressure drop through this valve.

What Special Equipment Should I Have?

If you are purchasing a high quality deaerator with high and low water alarms and pump low water cut-off, your surge tank should be equipped with the same alarms and cut-offs. In addition, since surge tanks are vented to atmosphere, they will corrode. Surge tanks should be cast iron, Plasite lined with #7156 or made of 300 series stainless steel for about the same price.

Determining Receiver Size

The receiver on a surge tank is sized based on the total load of all boilers in the system or the deaerator rating, the same sizing as that of a standard boiler feed unit. The receiver size is generally based on **10 minutes of net storage** as a general rule of thumb just like the deaerators.

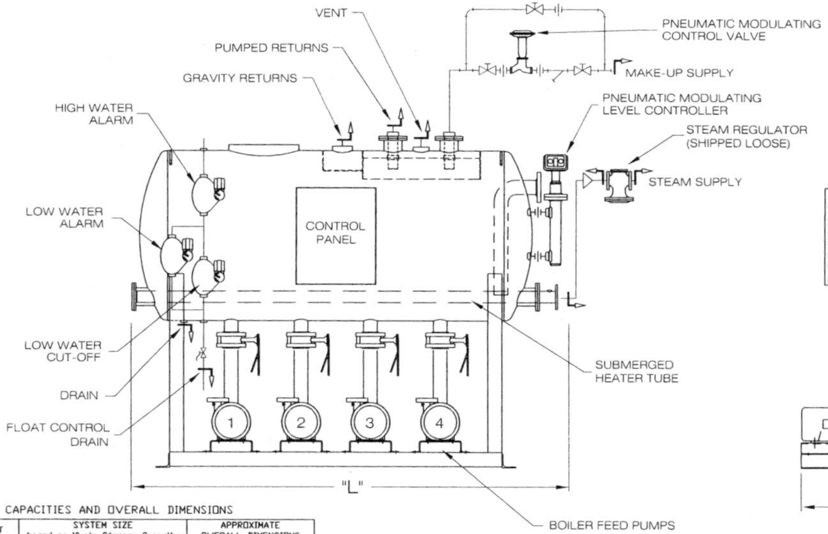

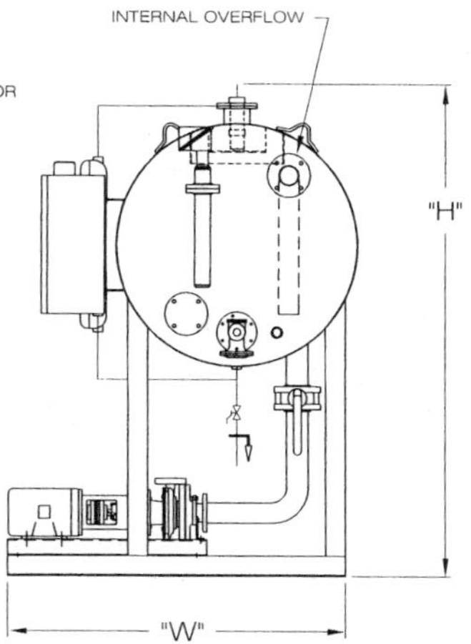

SELECTED CAPACITIES AND OVERALL DIMENSIONS

STANDARD RECEIVER SIZE	THICK	NET RECEIVER GALLONS	SYSTEM SIZE based on 10 min. Storage Capacity		APPROXIMATE OVERALL DIMENSIONS		
			Boiler Horsepower	lb/hr Steam	Length	Width	Height
36 × 60	3/16	215	290	10,000	102	94	100
36 × 72	3/16	257	350	12,000	114	94	100
36 × 84	3/16	293	435	15,000	126	94	100
42 × 72	3/16	380	550	19,000	116	98	102
42 × 84	3/16	438	640	22,000	128	98	102
42 × 96	3/16	482	695	24,000	140	98	102
48 × 84	1/4	538	780	27,000	130	100	110
48 × 96	1/4	594	870	30,000	142	100	110
48 × 120	1/4	715	1,045	36,000	166	100	110
54 × 120	5/16	987	1,450	50,000	170	104	116
60 × 120	5/16	1,273	1,855	64,000	172	106	124
60 × 144	5/16	1,479	2,145	74,000	196	106	124
66 × 120	5/16	1,641	2,375	82,000	174	110	130
66 × 144	5/16	1,907	2,755	95,000	198	110	130
72 × 120	3/8	2,036	2,955	102,000	174	122	136
72 × 144	3/8	2,375	3,480	120,000	198	122	136

Width dimension may vary based on pump model selection

Base footprint is approximate, if smaller footprint desired, consult factory for customization.

"H" DIMENSION BASED ON 48" ELEVATION

INDICATES WHERE SHIPCO PIPING ENDS

"ALL DIMENSIONS ARE APPROXIMATE"

.03 DA
.03 DA-2T
ATMOSPHERIC DESIGN

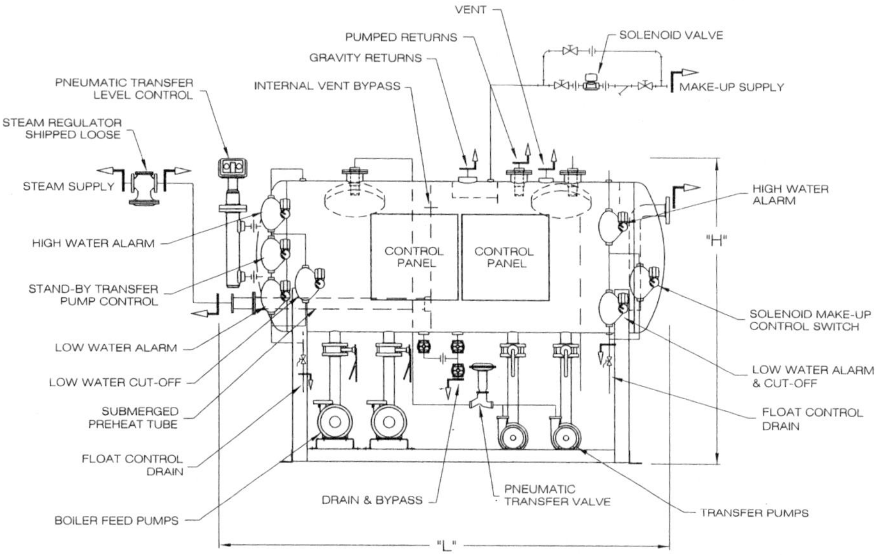

SELECTED CAPACITIES AND OVERALL DIMENSIONS

STANDARD RECEIVER SIZE	THICK	NET RECEIVER GALLONS	SYSTEM SIZE based on 10 min. Storage Capacity		APPROXIMATE OVERALL DIMENSIONS		
			Boiler Horsepower	lb/hr Steam	Length	Width	Height
36 x 60	3/16	230	145	5,000	102	94	100
36 x 72	3/16	257	175	6,000	114	94	100
36 x 84	3/16	293	220	7,500	126	94	100
42 x 72	3/16	380	275	9,500	116	98	102
42 x 84	3/16	438	320	11,000	128	98	102
42 x 96	3/16	482	350	12,000	140	98	102
48 x 84	1/4	538	390	13,500	130	100	110
48 x 96	1/4	594	435	15,000	142	100	110
48 x 120	1/4	715	520	18,000	166	100	110
54 x 120	5/16	987	725	25,000	170	104	116
60 x 120	5/16	1,273	930	32,000	172	106	124
60 x 144	5/16	1,479	1,070	37,000	196	106	124
66 x 120	5/16	1,641	1,190	41,000	124	110	130
66 x 144	5/16	1,907	1,375	47,500	198	110	130
72 x 120	3/8	2,036	1,480	51,000	174	122	136
72 x 144	3/8	2,375	1,740	60,000	198	122	136
72 x 168	3/8	2,675	1,940	67,000	222	122	136
72 x 192	3/8	3,000	2,175	75,000	246	122	136
72 x 204	3/8	3,250	2,350	81,000	258	122	136
84 x 144	3/8	3,365	2,440	84,000	202	130	148
84 x 168	3/8	3,828	2,775	95,000	226	130	148
84 x 192	3/8	4,285	3,100	107,000	250	130	148
96 x 204	3/8	5,686	4,115	142,000	266	136	160

Width dimension may vary based on pump model selection

Base footprint is approximate, if smaller footprint desired, consult factory for customization.

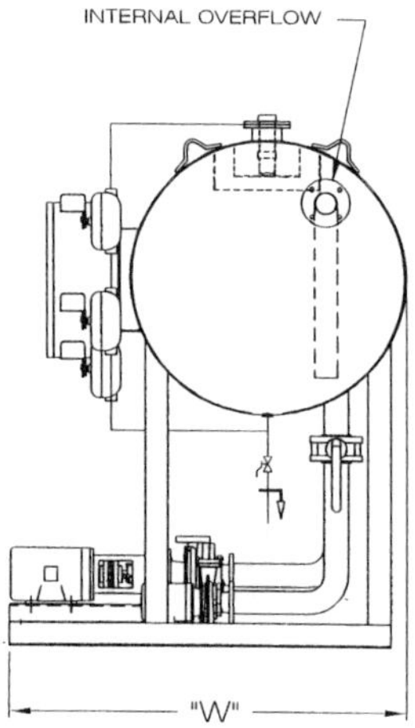

"H" DIMENSION BASED ON 48" ELEVATION

INDICATES WHERE SHIPCO PIPING ENDS

"ALL DIMENSIONS ARE APPROXIMATE"

.03 DA-2C
ATMOSPHERIC DESIGN

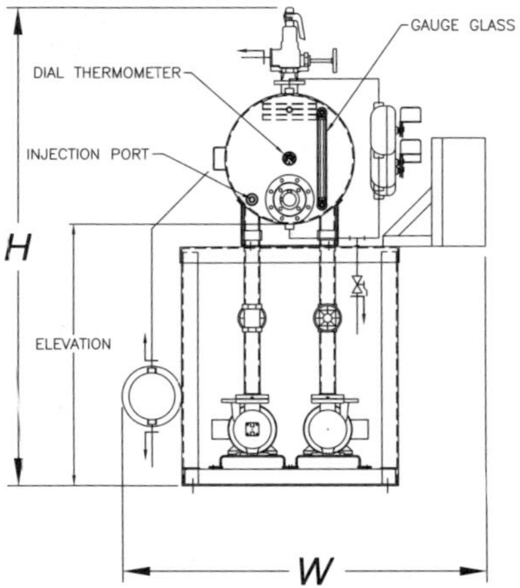

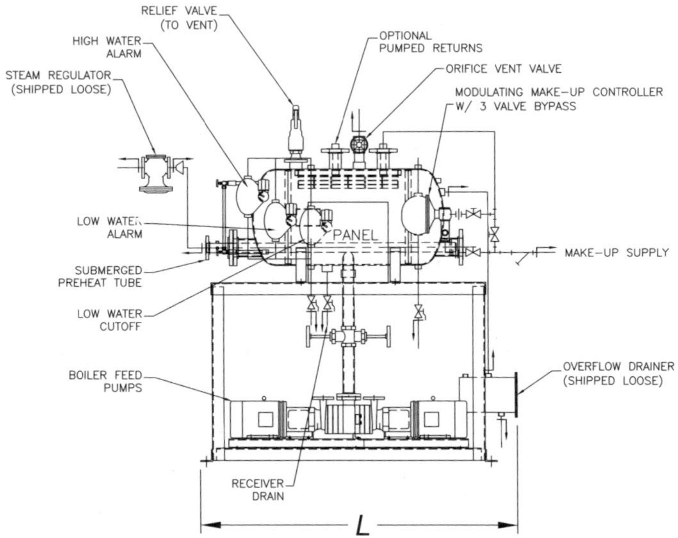

SELECTED CAPACITIES AND OVERALL DIMENSIONS

STANDARD RECEIVER SIZE	NET RECEIVER GALLONS	SYSTEM SIZE based on 10 min. Storage Capacity		APPROXIMATE OVERALL DIMENSIONS		
		Boiler Horsepower	lb/hr Steam	Length	Width	Height
24 x 24	36	52	1,795	74	68	88
24 x 30	42	60	2,070	74	68	88
24 x 36	48	70	2,415	74	68	88
24 x 48	60	87	3,000	80	68	88
24 x 60	72	104	3,590	86	68	88
30 x 36	85	123	4,245	74	72	94
30 x 48	110	160	5,520	80	72	94
30 x 60	135	196	6,760	86	72	94
36 x 48	160	232	8,005	80	76	100

Base footprint is approximate, if smaller footprint desired, consult factory for customization.

Based on 48" elevation

All dimensions are in inches

"H" DIMENSION BASED ON 48" ELEVATION

INDICATES WHERE SHIPCO PIPING ENDS

"ALL DIMENSIONS ARE APPROXIMATE"

.005 DA-SCP
PRESSURIZED DESIGN

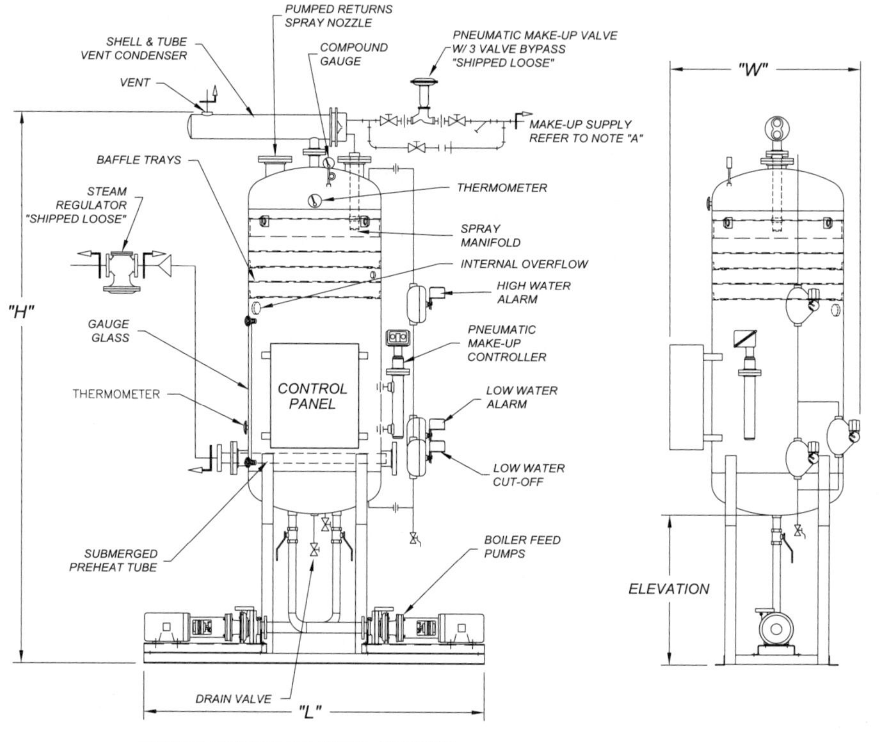

SELECTED CAPACITIES AND OVERALL DIMENSIONS

STANDARD RECEIVER SIZE	THICK	NET RECEIVER GALLONS	SYSTEM SIZE based on 10 min. Storage Capacity		APPROXIMATE OVERALL DIMENSIONS		
			Boiler Horsepower	lb/hr Steam	Length	Width	Height
24 x 48	10 gal.	38	55	2,000	50	50	132
24 x 60	10 gal.	60	90	3,000	50	50	144
30 x 60	10 gal.	95	145	5,000	56	56	147
36 x 60	10 gal.	132	205	7,000	62	62	150
42 x 60	10 gal.	180	260	9,000	68	68	155

Length dimension may vary based on pump model selection

Base footprint is approximate, if smaller footprint desired, consult factory for customization.

"H" DIMENSION BASED ON 48" ELEVATION

INDICATES WHERE SHIPCO PIPING ENDS

"ALL DIMENSIONS ARE APPROXIMATE"

NOTE "A"
MAKE-UP CONNECTION IF SINGLE TANK DESIGN
TRANSFER CONNECTION IF TWO TANK DESIGN

.005 DA-STV
.005 DA-STV-2T
ATMOSPHERIC DESIGN

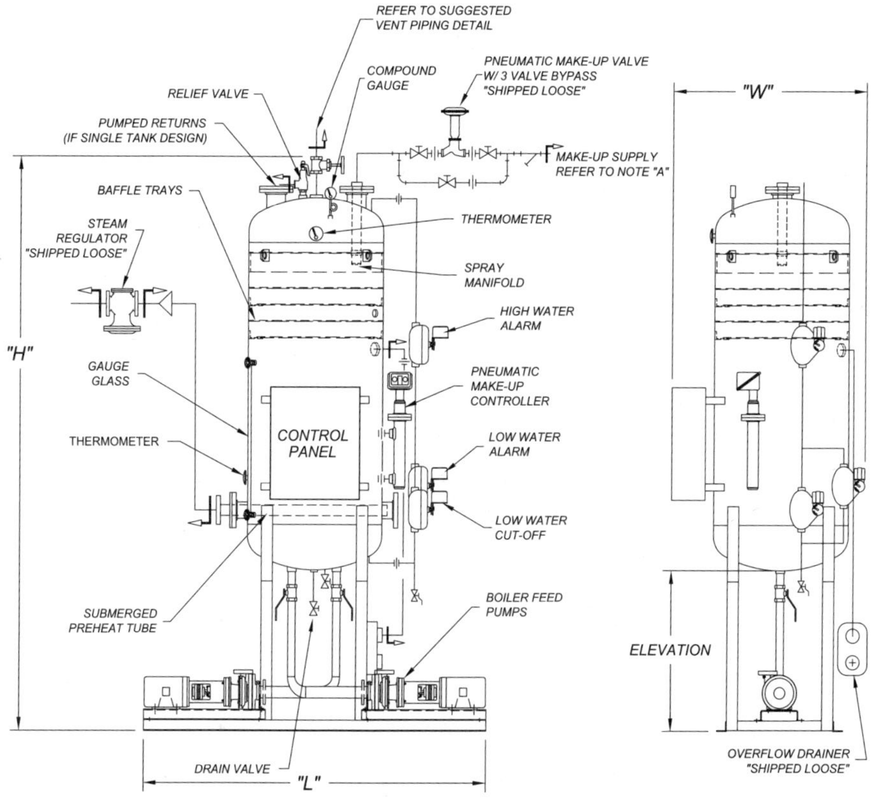

SELECTED CAPACITIES AND OVERALL DIMENSIONS

STANDARD RECEIVER SIZE	NET RECEIVER GALLONS	SYSTEM SIZE based on 10 min. Storage Capacity		APPROXIMATE OVERALL DIMENSIONS		
		Boiler Horsepower	lb/hr Steam	Length	Width	Height
24 x 48	38	55	2,000	50	50	134
24 x 60	60	90	3,000	50	50	146
30 x 60	95	145	5,000	56	56	150
36 x 60	132	205	7,000	62	62	152
42 x 60	180	260	9,000	68	68	160

Length dimension may vary based on pump model selection

Base footprint is approximate, if smaller footprint desired, consult factory for customization.

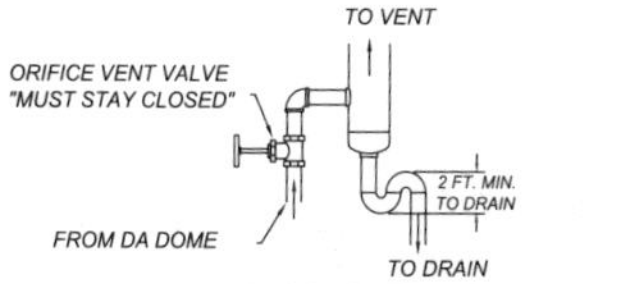

SUGGESTED VENT PIPING

NOTE "A"
MAKE-UP CONNECTION IF SINGLE TANK DESIGN
TRANSFER CONNECTION IF TWO TANK DESIGN

"H" DIMENSION BASED ON 48" ELEVATION
◁ INDICATES WHERE SHIPCO PIPING ENDS ▷
"ALL DIMENSIONS ARE APPROXIMATE"

.005 DA-STVP
.005 DA-STVP-2T
PRESSURIZED DESIGN

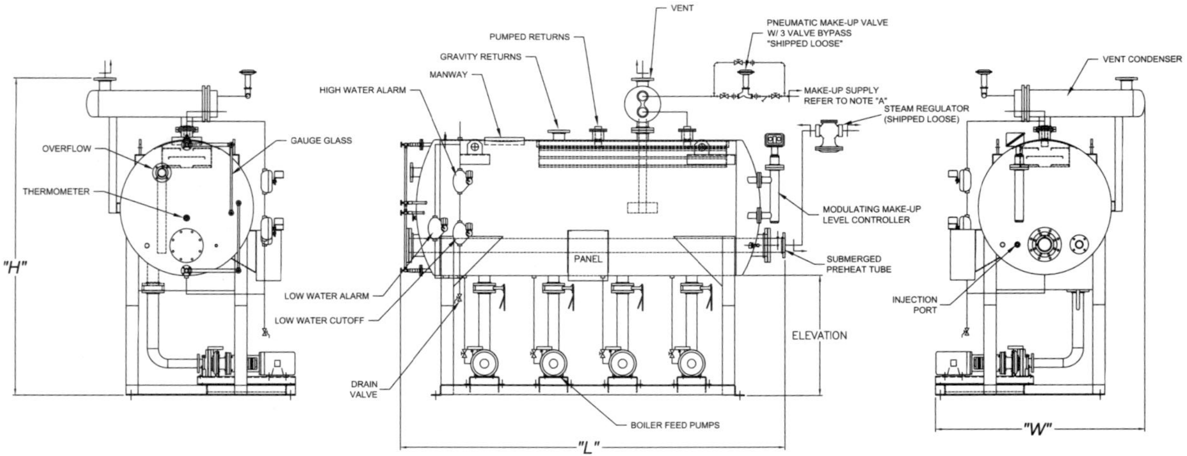

SELECTED CAPACITIES AND OVERALL DIMENSIONS

STANDARD RECEIVER SIZE	THICK	NET RECEIVER GALLONS	SYSTEM SIZE based on 10 min. Storage Capacity		APPROXIMATE OVERALL DIMENSIONS		
			Boiler Horsepower	lb/hr Steam	Length	Width	Height
36 x 60	3/16	180	260	9,000	102	90	104
36 x 72	3/16	223	320	11,000	114	90	104
36 x 84	3/16	260	375	13,000	126	90	104
42 x 72	3/16	306	435	15,000	116	93	110
42 x 84	3/16	358	520	18,000	128	93	110
42 x 96	3/16	400	580	20,000	140	93	110
48 x 84	1/4	465	665	23,000	130	96	116
48 x 96	1/4	532	780	27,000	142	96	116
48 x 120	1/4	665	985	34,000	166	96	116
54 x 120	5/16	825	1,215	42,000	170	100	122
60 x 120	5/16	1,025	1,505	52,000	172	102	128
60 x 144	5/16	1,295	1,885	65,000	196	102	128
66 x 120	5/16	1,440	2,085	72,000	174	105	134
66 x 144	5/16	1,670	2,435	84,000	198	105	134
72 x 120	3/8	1,785	2,610	90,000	174	118	140
72 x 144	3/8	2,080	3,015	104,000	198	118	140

Width dimension may vary based on pump model selection

Base footprint is approximate, if smaller footprint desired, consult factory for customization.

"H" DIMENSION BASED ON 48" ELEVATION

INDICATES WHERE SHIPCO PIPING ENDS

"ALL DIMENSIONS ARE APPROXIMATE"

NOTE "A"

MAKE-UP CONNECTION IF SINGLE TANK DESIGN
TRANSFER CONNECTION IF TWO TANK DESIGN

.005 DA-IST
.005 DA-IST-2T
ATMOSPHERIC DESIGN

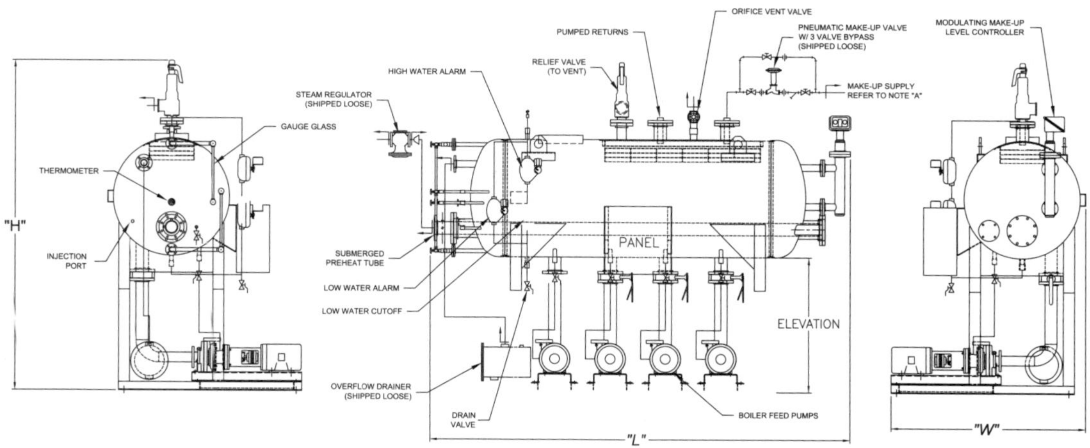

SELECTED CAPACITIES AND OVERALL DIMENSIONS

STANDARD RECEIVER SIZE	NET RECEIVER GALLONS	SYSTEM SIZE based on 10 min. Storage Capacity		APPROXIMATE OVERALL DIMENSIONS		
		Boiler Horsepower	lb/hr Steam	Length	Width	Height
36 x 60	180	260	9,000	108	80	106
36 x 72	223	320	11,000	120	80	106
36 x 84	260	375	13,000	132	80	106
42 x 72	306	435	15,000	124	83	118
42 x 84	358	520	18,000	136	83	118
42 x 96	400	580	20,000	148	83	118
48 x 84	465	665	23,000	139	86	124
48 x 96	532	780	27,000	151	86	124
48 x 120	665	985	34,000	175	86	124
54 x 120	825	1,215	42,000	177	90	136
60 x 120	1,025	1,505	52,000	180	92	142
60 x 144	1,295	1,885	65,000	205	92	142
66 x 120	1,440	2,085	72,000	184	95	148
66 x 144	1,670	2,435	84,000	208	95	148
72 x 120	1,785	2,610	90,000	186	108	154
72 x 144	2,080	3,015	104,000	210	108	154
72 x 168	2,340	3,390	117,000	234	108	154
72 x 192	2,625	3,795	131,000	258	108	154
72 x 204	2,845	4,145	143,000	270	108	154

Width dimension may vary based on pump model selection

Base footprint is approximate, if smaller footprint desirec, consult factory for customization.

"H" DIMENSION BASED ON 48" ELEVATION

INDICATES WHERE SHIPCO PIPING ENDS

"ALL DIMENSIONS ARE APPROXIMATE"

NOTE "A"
MAKE-UP CONNECTION IF SINGLE TANK DESIGN
TRANSFER CONNECTION IF TWO TANK DESIGN

.005 DA-ISTP
.005 DA-ISTP-2T
PRESSURIZED DESIGN

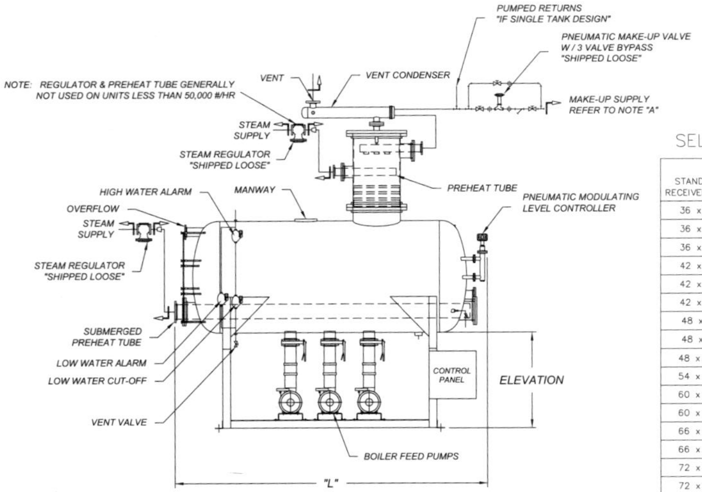

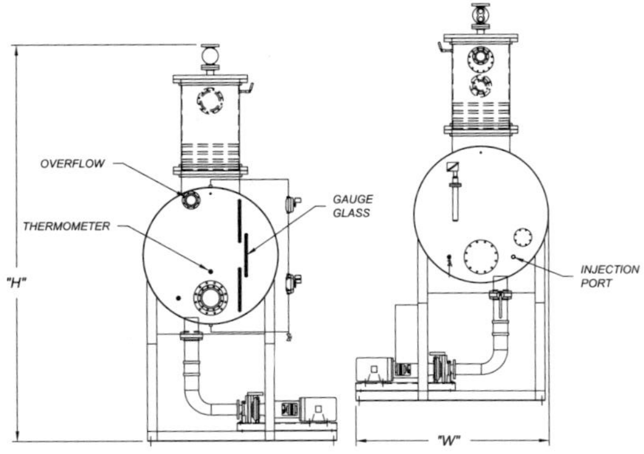

SELECTED CAPACITIES AND OVERALL DIMENSIONS

STANDARD RECEIVER SIZE	THICK	NET RECEIVER GALLONS	SYSTEM SIZE based on 10 min. Storage Capacity		APPROXIMATE OVERALL DIMENSIONS		
			Boiler Horsepower	lb/hr Steam	Length	Width	Height
36 x 60	3/16	230	350	12,000	112	80	166
36 x 72	3/16	257	375	13,000	122	80	166
36 x 84	3/16	293	435	15,000	134	80	166
42 x 72	3/16	380	580	20,000	126	83	172
42 x 84	3/16	438	640	22,000	138	83	172
42 x 96	3/16	482	695	24,000	150	83	172
48 x 84	1/4	538	810	28,000	142	86	178
48 x 96	1/4	594	870	30,000	154	86	178
48 x 120	1/4	715	1,045	36,000	178	86	178
54 x 120	5/16	987	1,450	50,000	182	90	200
60 x 120	5/16	1,273	1,855	64,000	184	92	206
60 x 144	5/16	1,479	2,145	74,000	210	92	206
66 x 120	5/16	1,641	2,375	82,000	188	95	214
66 x 144	5/16	1,907	2,755	95,000	210	95	214
72 x 120	3/8	2,036	2,955	102,000	190	108	240
72 x 144	3/8	2,375	3,480	120,000	214	108	240
72 x 168	3/8	2,675	3,885	134,000	240	108	240
72 x 192	3/8	3,000	4,350	150,000	262	108	240
72 x 204	3/8	3,250	4,785	165,000	274	108	240
84 x 132	3/8	3,134	4,540	157,000	216	114	250
84 x 144	3/8	3,365	4,875	168,000	220	114	250
84 x 168	3/8	3,828	5,550	190,000	242	114	250
84 x 192	3/8	4,285	6,200	215,000	268	114	250
96 x 204	3/8	5,686	8,260	285,000	284	120	270

Width dimension may vary based on pump model selection

Height dimension based on 48" elevation

Base footprint is approximate, if smaller footprint desired, consult factory for customization.

"H" DIMENSION BASED ON 48" ELEVATION

INDICATES WHERE SHIPCO PIPING ENDS

"ALL DIMENSIONS ARE APPROXIMATE"

NOTE "A"
MAKE-UP CONNECTION IF SINGLE TANK DESIGN
TRANSFER CONNECTION IF TWO TANK DESIGN

.005 DA-ST
.005 DA-ST-2T
ATMOSPHERIC DESIGN

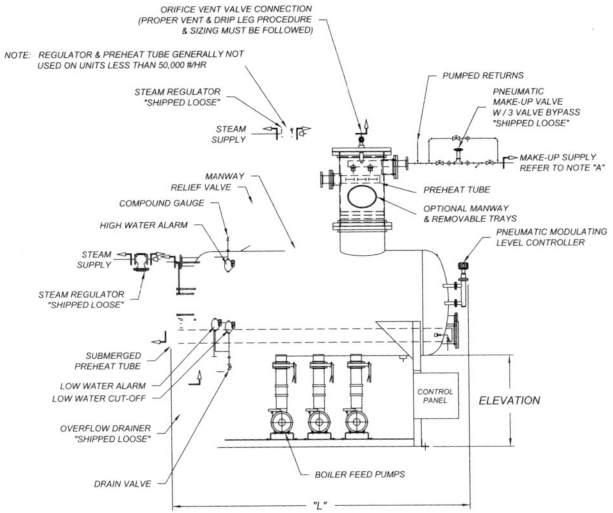

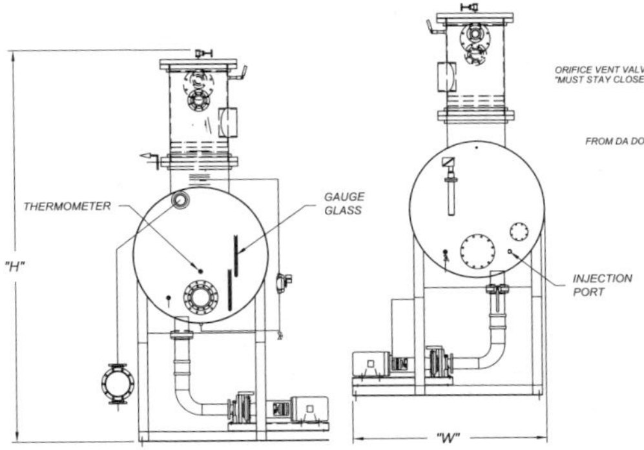

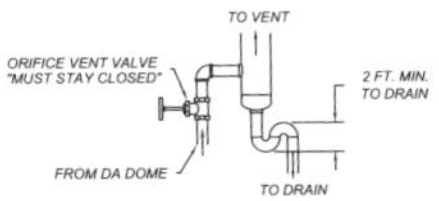

SUGGESTED VENT PIPING

SELECTED CAPACITIES AND OVERALL DIMENSIONS

STANDARD RECEIVER SIZE	NET RECEIVER GALLONS	SYSTEM SIZE based on 10 min. Storage Capacity		APPROXIMATE OVERALL DIMENSIONS		
		Boiler Horsepower	lb/hr Steam	Length	Width	Height
36 x 60	230	350	12,000	110	80	156
36 x 72	257	375	13,000	122	80	156
36 x 84	293	435	15,000	134	80	156
42 x 72	380	580	20,000	126	83	170
42 x 84	438	640	22,000	138	83	170
42 x 96	482	695	24,000	150	83	170
48 x 84	538	810	28,000	142	86	176
48 x 96	594	870	30,000	154	86	176
48 x 120	715	1,045	36,000	178	86	176
54 x 120	987	1,450	50,000	180	90	190
60 x 120	1,273	1,855	64,000	184	92	196
60 x 144	1,479	2,145	74,000	208	92	196
66 x 120	1,641	2,375	82,000	186	95	204
66 x 144	1,907	2,755	95,000	210	95	204
72 x 120	2,036	2,955	102,000	190	108	235
72 x 144	2,375	3,480	120,000	214	108	235
72 x 168	2,675	3,885	134,000	238	108	235
72 x 192	3,000	4,350	150,000	262	108	235
72 x 204	3,250	4,785	165,000	274	108	235
84 x 132	3,134	4,540	157,000	204	114	260
84 x 144	3,365	4,875	168,000	216	114	260
84 x 168	3,828	5,550	190,000	240	114	260
84 x 192	4,285	6,200	215,000	264	114	260
96 x 204	5,686	8,260	285,000	280	120	270
108 x 168	6,446	9,420	325,000	254	126	285
108 x 180	6,830	9,855	340,000	266	126	285
108 x 198	7,396	10,725	370,000	284	126	285
108 x 216	7,986	11,595	400,000	302	126	285
108 x 264	9,430	14,495	500,000	350	126	285

Width dimension may vary based on pump model selection

Height dimension based on 48" elevation

Base footprint is approximate, if smaller footprint desired, consult factory for customization.

"H" DIMENSION BASED ON 48" ELEVATION

INDICATES WHERE SHIPCO PIPING ENDS

"ALL DIMENSIONS ARE APPROXIMATE"

NOTE "A"
MAKE-UP CONNECTION IF SINGLE TANK DESIGN
TRANSFER CONNECTION IF TWO TANK DESIGN

.005 DA-STP
.005 DA-STP-2T
PRESSURIZED DESIGN

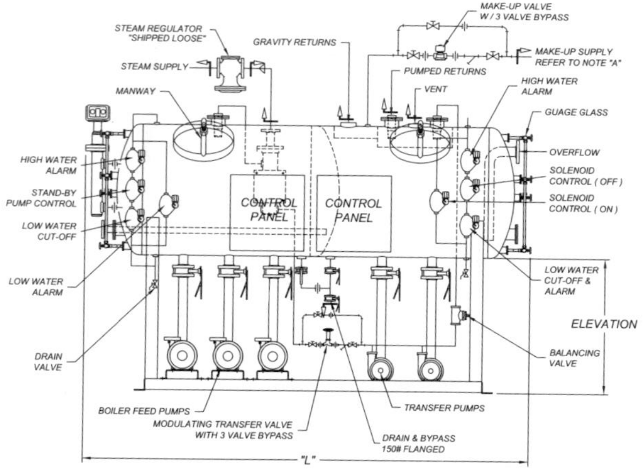

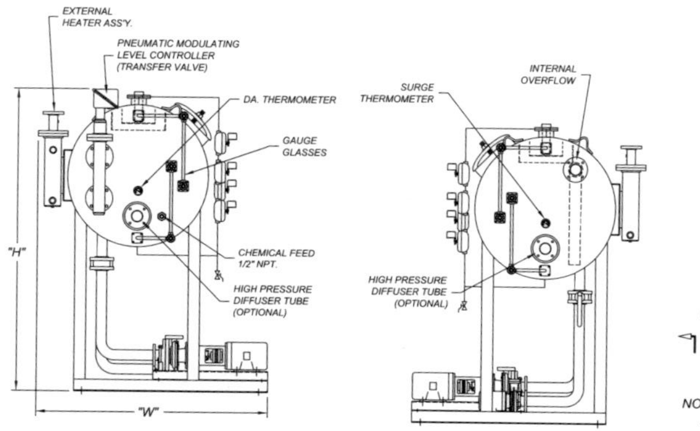

SELECTED CAPACITIES AND OVERALL DIMENSIONS

STANDARD RECEIVER SIZE	THICK	NET RECEIVER GALLONS	SYSTEM SIZE based on 10 min. Storage Capacity		APPROXIMATE OVERALL DIMENSIONS		
			Boiler Horsepower	lb/hr Steam	Length	Width	Height
36 x 60	3/16	230	145	5,000	102	94	100
36 x 72	3/16	257	175	6,000	114	94	100
36 x 84	3/16	293	220	7,500	126	94	100
42 x 72	3/16	380	275	9,500	116	98	102
42 x 84	3/16	438	320	11,000	128	98	102
42 x 96	3/16	482	350	12,000	140	98	102
48 x 84	1/4	538	390	13,500	130	100	110
48 x 96	1/4	594	435	15,000	142	100	110
48 x 120	1/4	715	520	18,000	166	100	110
54 x 120	5/16	987	725	25,000	170	104	116
60 x 120	5/16	1,273	930	32,000	172	106	124
60 x 144	5/16	1,479	1,070	37,000	196	106	124
66 x 120	5/16	1,641	1,190	41,000	174	110	130
66 x 144	5/16	1,907	1,360	47,000	198	110	130
72 x 120	3/8	2,036	1,480	51,000	174	122	136
72 x 144	3/8	2,375	1,740	60,000	198	122	136
72 x 168	3/8	2,675	1,940	67,000	222	122	136
72 x 192	3/8	3,000	2,175	75,000	246	122	136
72 x 204	3/8	3,250	2,350	81,000	258	122	136
84 x 144	3/8	3,365	2,440	84,000	202	130	148
84 x 168	3/8	3,828	2,775	95,000	226	130	148
84 x 192	3/8	4,285	3,100	107,000	250	130	148
96 x 204	3/8	5,686	4,115	142,000	266	136	160
108 x 168	7/16	6,446	4,695	162,000	240	140	172
108 x 180	7/16	6,830	4,955	171,000	252	140	172
108 x 198	7/16	7,396	5,360	185,000	270	140	172
108 x 216	7/16	7,986	5,795	200,000	288	140	172
108 x 264	7/16	9,430	7,245	250,000	336	140	172

Width dimension may vary based on pump model selection

Height dimension based on 48" elevation

Base footprint is approximate, if smaller footprint desired, consult factory for customization.

"H" DIMENSION BASED ON 48" ELEVATION

INDICATES WHERE SHIPCO PIPING ENDS

"ALL DIMENSIONS ARE APPROXIMATE"

NOTE "A"

MAKE-UP CONNECTION IF SINGLE TANK DESIGN

TRANSFER CONNECTION IF TWO TANK DESIGN

.005 DA-2C
ATMOSPHERIC DESIGN

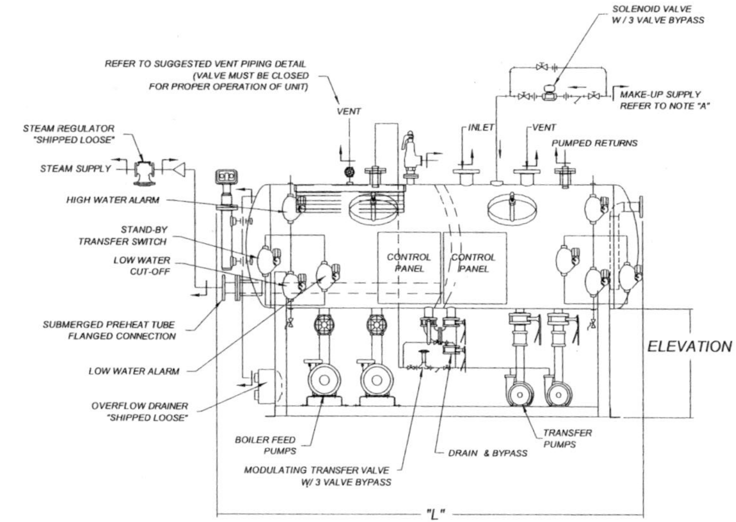

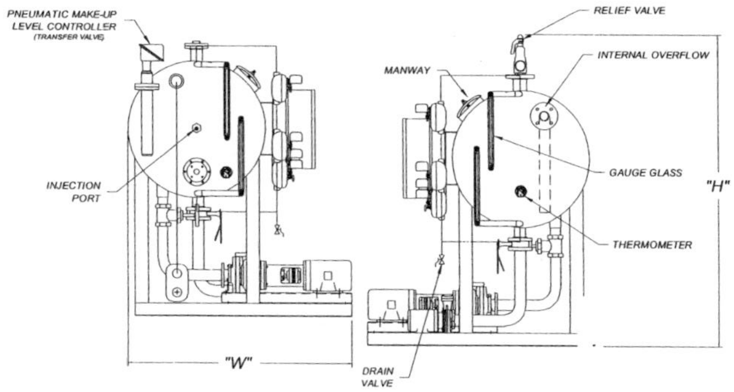

SELECTED CAPACITIES AND OVERALL DIMENSIONS

STANDARD RECEIVER SIZE	NET RECEIVER GALLONS	SYSTEM SIZE based on 10 min. Storage Capacity		APPROXIMATE OVERALL DIMENSIONS		
		Boiler Horsepower	lb/hr Steam	Length	Width	Height
36 x 60	220	145	5,000	110	80	110
36 x 72	244	175	6,000	120	80	110
38 x 84	278	205	7,000	134	80	110
42 x 72	361	260	9,000	126	83	116
42 x 84	416	290	10,000	138	83	116
42 x 96	458	335	11,500	150	83	116
48 x 84	511	375	13,000	142	86	130
48 x 96	565	405	14,000	154	86	130
48 x 120	680	495	17,000	178	86	130
54 x 120	938	680	23,500	180	90	136
60 x 120	1,210	870	30,000	184	92	142
60 x 144	1,405	1,015	35,000	208	92	142
66 x 120	1,560	1,130	39,000	186	95	148
66 x 144	1,812	1,305	45,000	210	95	148
72 x 120	1,934	1,390	48,000	190	108	154
72 x 144	2,256	1,640	56,500	214	108	154
72 x 168	2,542	1,840	63,500	238	108	154
72 x 192	2,850	2,060	71,000	262	108	154
72 x 204	3,088	2,260	78,000	274	108	154
84 x 144	3,200	2,300	80,000	216	114	166
84 x 168	3,640	2,600	90,000	240	114	166
84 x 192	4,070	2,950	102,000	264	114	166
96 x 204	5,402	3,915	135,000	280	120	180
108 x 168	6,045	4,495	155,000	254	126	196
108 x 180	6,405	4,785	165,000	266	126	196
108 x 198	6,940	5,130	177,000	284	126	196
108 x 216	7,475	5,505	190,000	302	126	196
108 x 264	8,900	6,955	240,000	350	126	196

Width dimension may vary based on pump model selection

Base footprint is approximate, if smaller footprint desired, consult factory for customization.

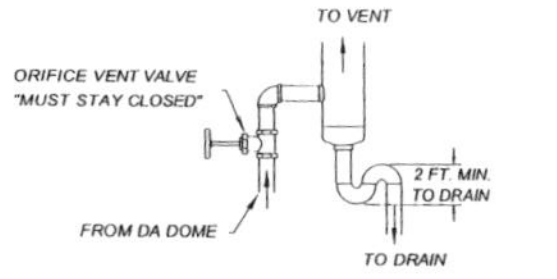

SUGGESTED VENT PIPING

"H" DIMENSION BASED ON 48" ELEVATION

INDICATES WHERE SHIPCO PIPING ENDS

"ALL DIMENSIONS ARE APPROXIMATE"

.005 DA-ISTP-2C PRESSURIZED DESIGN

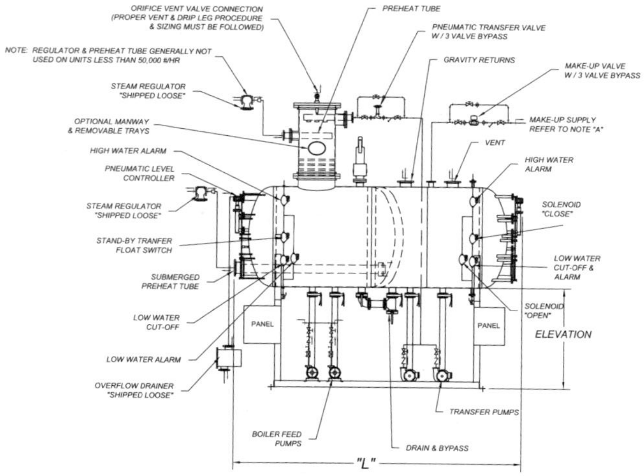

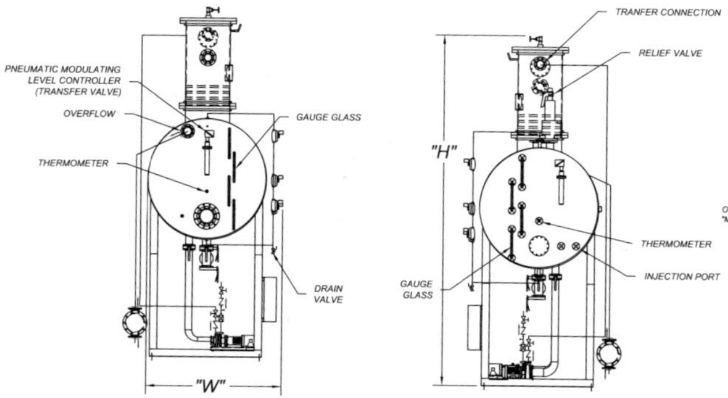

SELECTED CAPACITIES AND OVERALL DIMENSIONS

STANDARD RECEIVER SIZE	NET RECEIVER GALLONS	SYSTEM SIZE based on 10 min. Storage Capacity		APPROXIMATE OVERALL DIMENSIONS		
		Boiler Horsepower	lb/hr Steam	Length	Width	Height
36 x 60	230	145	5,000	110	80	156
36 x 72	257	175	6,000	122	80	156
38 x 84	293	220	7,500	134	80	156
42 x 72	380	275	9,500	126	83	170
42 x 84	438	320	11,000	138	83	170
42 x 96	482	350	12,000	150	83	170
48 x 84	538	390	13,500	142	86	176
48 x 96	594	435	15,000	154	86	176
48 x 120	715	520	18,000	178	86	176
54 x 120	987	725	25,000	180	90	190
60 x 120	1,273	930	32,000	184	92	196
60 x 144	1,479	1,070	37,000	208	92	196
66 x 120	1,641	1,190	41,000	186	95	204
66 x 144	1,907	1,360	47,000	210	95	204
72 x 120	2,036	1,480	51,000	190	108	235
72 x 144	2,375	1,740	60,000	214	108	235
72 x 168	2,675	1,940	67,000	238	108	235
72 x 192	3,000	2,175	75,000	262	108	235
72 x 204	3,250	2,350	81,000	274	108	235
84 x 144	3,365	2,440	84,000	216	114	260
84 x 168	3,828	2,775	95,000	264	114	260
84 x 192	4,285	3,100	107,000	264	114	260
96 x 204	5,686	4,115	142,000	280	120	270
108 x 168	6,446	4,695	162,000	254	126	285
108 x 180	6,830	4,955	171,000	266	126	285
108 x 198	7,396	5,360	185,000	284	126	285
108 x 216	7,986	5,795	200,000	302	126	285
108 x 264	9,430	7,245	250,000	350	126	285

Width dimension may vary based on pump model selection

Base footprint is approximate, if smaller footprint desired, consult factory for customization.

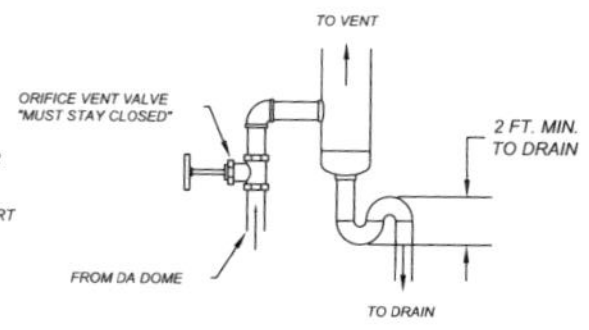

SUGGESTED VENT PIPING

"H" DIMENSION BASED ON 48" ELEVATION

INDICATES WHERE SHIPCO PIPING ENDS

"ALL DIMENSIONS ARE APPROXIMATE"

NOTE "A"
MAKE-UP CONNECTION IF SINGLE TANK DESIGN
TRANSFER CONNECTION IF TWO TANK DESIGN

.005 DA-STP-2C PRESSURIZED DESIGN

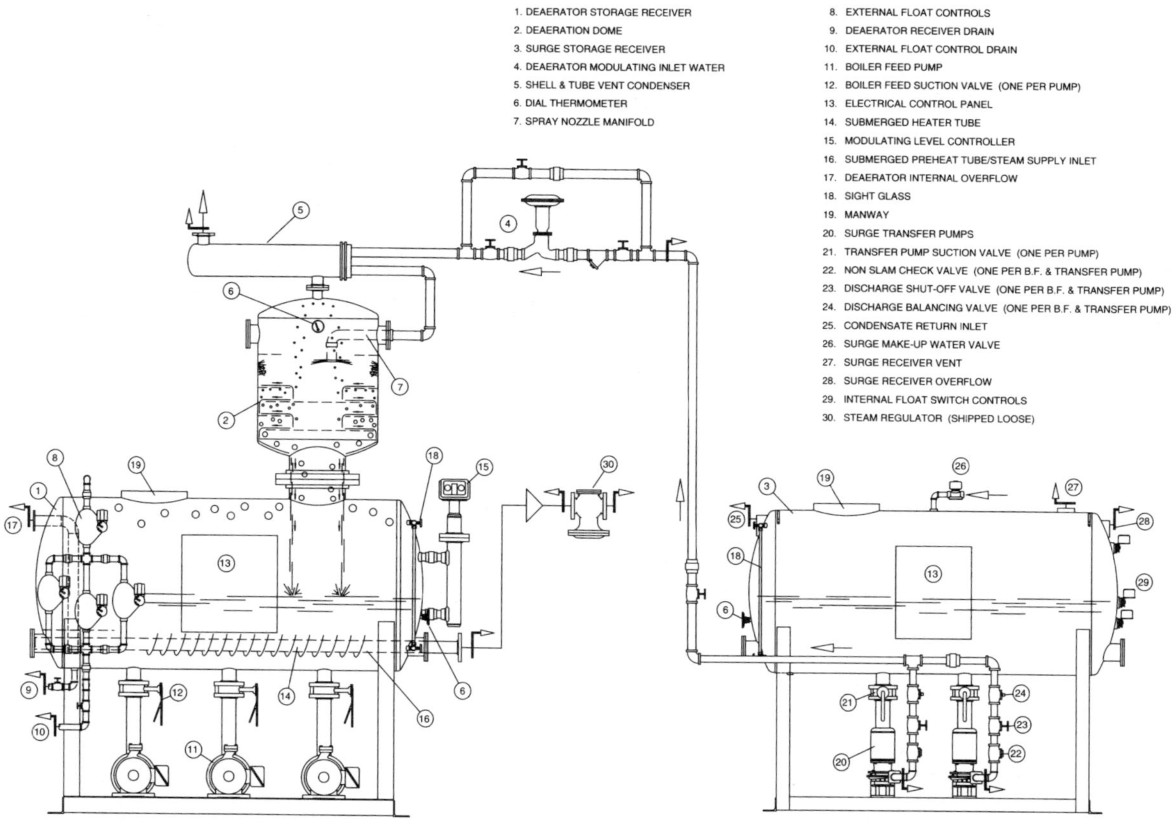

INDICATES WHERE SHIPCO PIPING ENDS

PRESSURIZED DA
WITH SURGE TANK

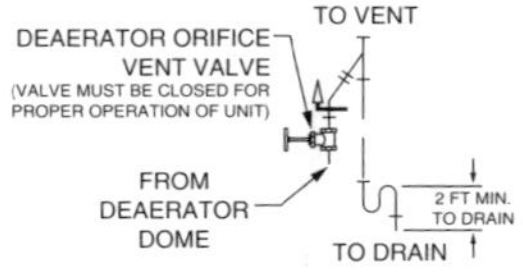

SUGGESTED VENT PIPING DETAIL

1. DEAERATOR STORAGE RECEIVER
2. DEAERATION DOME
3. SURGE STORAGE RECEIVER
4. DEAERATOR MODULATING INLET WATER
5. PRESSURE RELIEF VALVE
6. DEAERATOR ORIFICE VENT VALVE
7. DIAL THERMOMETER
8. SPRAY NOZZLE MANIFOLD
9. EXTERNAL FLOAT CONTROLS
10. DEAERATOR RECEIVER DRAIN

11. EXTERNAL FLOAT CONTROL DRAIN
12. BOILER FEED PUMPS
13. BOILER FEED SUCTION VALVE (ONE PER PUMP)
14. STEAM REGULATOR (SHIPPED LOOSE)
15. SUBMERGED HEATER TUBE
16. MODULATING LEVEL CONTROLLER
17. SUBMERGED PREHEAT TUBE/STEAM SUPPLY INLET
18. OVERFLOW DRAINER (SHIPPED LOOSE)
19. SIGHT GLASS
20. DIAL THERMOMETER
21. SURGE TRANSFER PUMPS
22. TRANSFER PUMP SUCTION VALVE (ONE PER PUMP)
23. NON SLAM CHECK VALVE (ONE PER B.F. & TRANSFER PUMP)
24. DISCHARGE SHUT-OFF VALVE (ONE PER B.F. & TRANSFER PUMP)
25. DISCHARGE BALANCING VALVE (ONE PER B.F. & TRANSFER PUMP)
26. CONDENSATE RETURN INLET
27. SURGE MAKE-UP WATER VALVE
28. SURGE RECEIVER VENT
29. SURGE RECEIVER OVERFLOW
30. FLOAT SWITCH CONTROLS
31. SIGHT GLASS
32. DIAL THERMOMETER
33. MANWAY

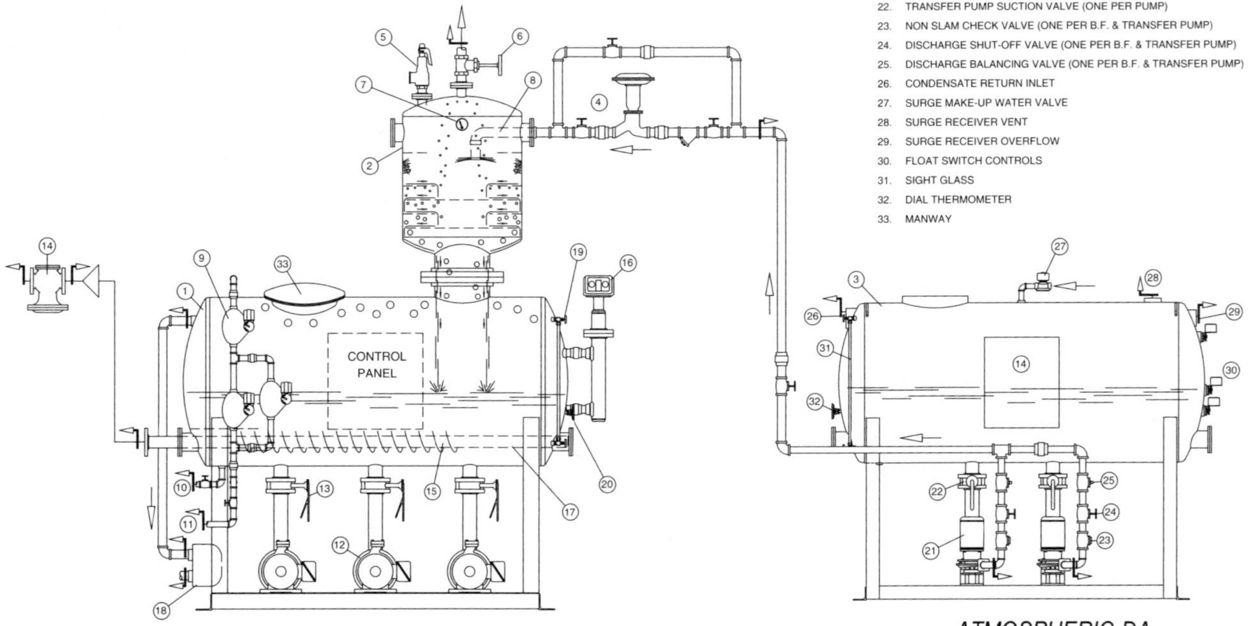

INDICATES WHERE SHIPCO PIPING ENDS

ATMOSPHERIC DA WITH SURGE TANK

Where Can You Go Wrong?

Following are problems that may occur when the deaerator is used under conditions not intended by its design, or when conditions arise that have not been previously considered:

1) Pressure errors or variations

   The actual **operating** boiler pressure is the design criterion, not necessarily the boiler design pressure. Often the actual pressure is substantially lower, as when some higher pressure boilers go to a low pressure night setback. This setback drastically affects pumps and steam control valves. In some cases only a reduced pressure is available to the steam control valve.

2) Temperature and capacity variations.

   A surge tank can greatly minimize temperature and capacity variations to the deaeration chamber. Any flow into the surge tank, whether hot pumped returns or cold make-up water, only slightly affects the temperature of the stored volume of water in the tank. The modulated flow out of the surge tank then minimizes capacity variations to the deaeration chamber.

   Boiler blowdown can greatly increase capacity for a short time; if this brings in cold make-up water, temperatures can change drastically. In a single compartment deaerator, this sometimes large transient can reduce temperatures, hurting deaeration, and impose short time loads on the deaerator beyond the intended design.

   If the actual amount of make-up water is higher than anticipated, or if return temperatures are lower, it can substantially affect the sizing of the steam control valve, steam manifold, and other parts.

If a large condensate pump starts pumping to a single compartment deaerator, it can substantially increase the steam required for the duration of the cycle.

3) Night pressure setback or night shut-offs.

   Night shut-off is self-defeating, as a lot of gases will dissolve overnight in the deaerator water as well as the boiler water, and two to six hours would then be required to restart the deaerator and purge the deaerator and boiler. The best solution is to isolate the boiler and deaerator from the system at night, improve the insulation, and run it on night setback.

   If a high pressure boiler is to be run on night setback at about 15 PSI, low pressure pumps with appropriate controls should be used at night. A lower pressure steam valve is needed when operating on low pressure pumps.

4) If boiler loads get very light, continuously running pumps on moderate HP can generate more heat in the pump than the small flow of water can carry away. Steam is then generated in the impeller, and the pumping of water ceases. Therefore consult factory for sizing.

5) Because air lends itself better to modulating controls than electricity, pneumatic controls are our standard. It is therefore important to ascertain whether the installation will have an air supply. If not, electrical controls can be furnished.

6) A deaerator requires steam for heating, and sometimes uncontrolled flows are returned to the

deaerator. This is acceptable for drip lines or very small flows; larger uncontrolled flows should go to a flash tank instead, as the steam to the deaerator must be carefully controlled to obtain good deaeration.

7) Returns can flow by gravity into an atmospheric deaerator, but not into a single-compartment pressurized deaerator. If gravity returns are to enter a pressurized deaerator, they must first go to a condensate pump and be pumped into the deaerator with other pumped returns.

8) Deaerators are usually specified for the maximum continuous load, but if condensate pumps are off and then on at a higher than normal flow, or if boiler blowdown results in high peak flows, these peaks must be acknowledged and considered in the design to obtain proper operation. The best solution is a surge tank with modulated output to smooth out these peak loads and provide reliable control of the deaerator.

9) Oversizing the storage in the deaerator tank to avoid using a surge tank. A deaerator is designed to maintain a constant level of water inside and good deaeration is achieved by controlling the wide temperature variations.

10) Oversizing the deaerator rating will lead to your regulator and modulating valves being oversized and lead them to wire draw. In addition on a pressurized deaerator you lose ½ of 1 percent of the load out the vent, and if you increase rating you lose more steam. Finally, the motor HP required will be more than you actually need, causing the customer to lose more money.

11) Using a lining to protect the steel receiver where the deaerated water lies is a waste of money. By lining this section in your deaerator, somebody is admitting that the deaerator doesn't work. When deaeration levels of .005 are present in the unit very little corrosion will take place.

12) Failure to put automatic flow control valves or balancing valves on the discharge of centrifugal boiler feed pumps. The centrifugal pumps discharge pressure must equal what the pump's impeller was trimmed for at the factory. Failure to control the pressure and flow will cause the pump's NPSH characteristics to go up and cause the pump's performance to diminish or stop.

13) Surge tank pumps are to run constantly and pump into the modulating transfer valve on the deaerator unit. Commonly, people want to run these pumps on–off, which in turn makes it no different than a condensate pump providing large temperature and capacity variations.

14) On a two-tank or two-compartment system, make-up water is to be added into the surge chamber and mixed with returns. If make-up water is directly added into the deaerator, every time the unit needs water the temperature will shock the deaerator and cause the regulator to open faster. This shock is caused by the wide temperature variation of having the regulator respond to a temperature of 50°F when it is used to 150°F.

15) Suction strainers are never used with centrifugal boiler feed pumps. A suction strainer will cause a pump to destroy itself by doing its job of collecting dirt and debris. No one can calculate

the pressure drop through a suction strainer and therefore you can't calculate stand height to avoid NPSH problems. The strainers belong in the return lines and make-up lines to the unit. The reason suction strainers are commonly found is that they are used with turbine pumps. Turbine pumps were the forerunner to today's multi-stage centrifugals, and these turbines had to have suction strainers because of the close tolerance and wear. A little bit of dirt could and would destroy these pumps, causing high maintenance problems. Centrifugal pumps are much more durable and can handle the dirt and debris.

16) In hospital applications a two-tank system is better than a two-compartment or duo-tank design. Hospitals can't go down, and when two compartment units need replaced the customer must get a rental unit. It would be smarter to put a two-tank system in so you could have the surge tank pumps feed directly into the suction of the boiler feed pumps and inspect or replace the deaerator. If you want the surge tank replaced or to be inspected, run the make-up directly into the deaerator and bypass the surge.

.03 DA-2C

.03 DA-CI

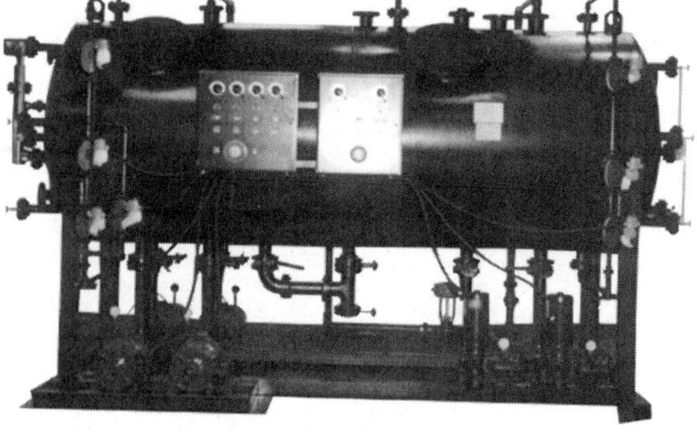

.005 DA-2C

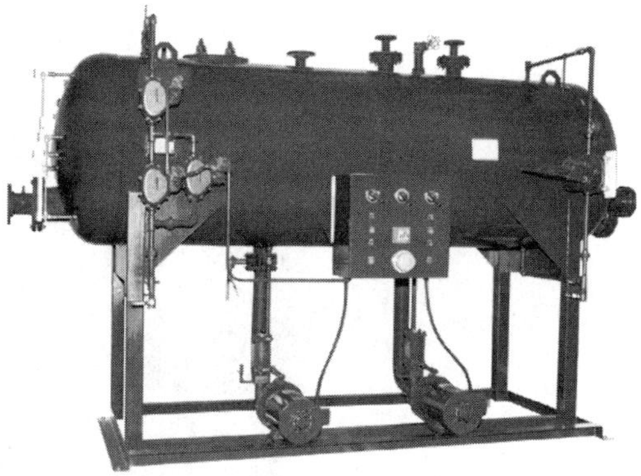

.005 DA-ISTP

.005 DA-ST

.005 DA-STP-2C

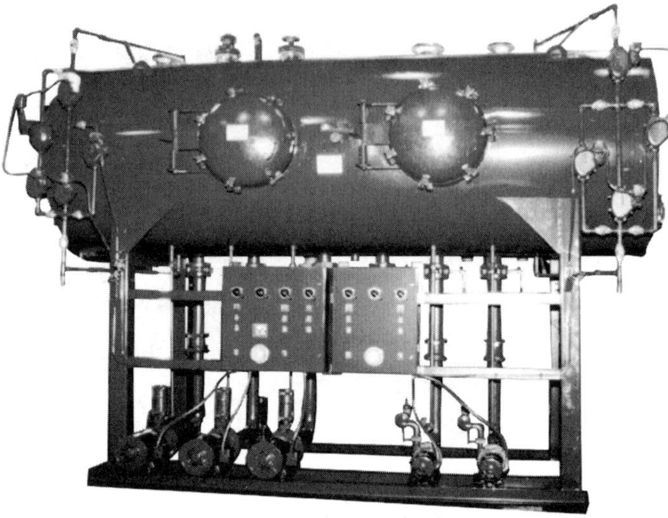

.005 DA-ISTP-2C

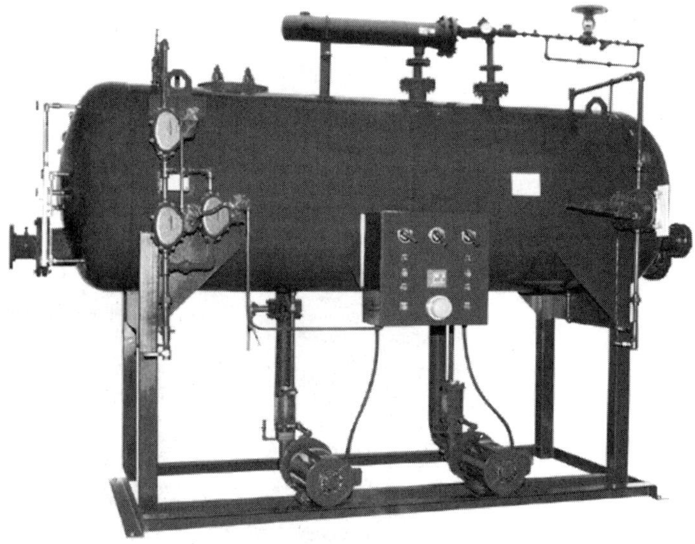

.005 DA-IST

.005 DA-STP

.005 DA-STVP

Pump and Motor Assembly

Type AWF